AVIATION MARK

To the memory of my parents

Aviation Markets
Studies in Competition and Regulatory Reform

DAVID STARKIE

ASHGATE

Published by
Ashgate Publishing Limited
Gower House
Croft Road
Aldershot
Hampshire GU11 3HR
England

Ashgate Publishing Company
Suite 420
101 Cherry Street
Burlington, VT 05401-4405
USA

Ashgate website: http://www.ashgate.com

British Library Cataloguing in Publication Data
Starkie, D. N. M. (David Nicholas Martin), 1942-
   Aviation markets : studies in competition and regulatory
   reform
   1. Aeronautics and state 2. Airlines - Deregulation - Case
   studies 3. Aeronautics, Commercial - Deregulation - Case
   studies
   I. Title
   387.7

Library of Congress Cataloging-in-Publication Data
Starkie, D. N. M. (David Nicholas Martin), 1942-
   Aviation markets : studies in competition and regulatory reform / by David Starkie.
      p. cm.
Includes bibliographical references and index.
   ISBN: 978-0-7546-7360-6 (hbk) -- ISBN: 978-0-7546-7388-0 (pbk)
  1. Aeronautics and state. 2. Airlines--Deregulation--Case studies. 3. Aeronautics,
Commercial--Deregulation--Case studies. I. Title.
   HE9777.7.S73 2008
   387.7--dc22

2007042388

ISBN: 978-0-7546-7360-6 (HBK)
ISBN: 978-0-7546-7388-0 (PBK)

Printed and bound in Great Britain by TJ International Ltd, Padstow, Cornwall.

Contents

List of Boxes, Figures and Tables

Boxes

Figures

Tables

Preface

Little more than 25 years ago, the global aviation industry was dominated by national and local governments and, that part of it that was not public sector controlled and operated, was highly regulated. In less than a generation this situation has changed radically. The industry, in large measure, is now placed within the market economy: ownership of large parts has passed into private hands, competitive forces determine or at least influence outcomes in many sectors and regions and, in turn, enterprise has flourished with the development of new business models and the introduction to the market of many new product lines. This, as studies have shown, has been of enormous benefit to the consumer and the global economy in general.

Within this broad canvas of change, two notable developments in public policy have acted as catalysts. First, the decision made in the late 1970s to liberalise the US domestic airline industry, then in private hands but highly regulated. The consequences of this public policy initiative were such that aviation policy in Europe and Australasia soon followed the US example. The second policy development of note was the decision in the mid-1980s to privatise the state-owned British Airports Authority which at the time controlled around three-quarters of the UK airport industry (measured by output). This privatisation was one of a series of similar sales of state-owned public utilities launched by the government of Margaret Thatcher. The vehicle used for selling these industries, including the British Airports Authority, was a public flotation with the result that the world had its first publicly quoted airport company and one with a substantial market capitalisation.

But if, as a result of these and related developments, world aviation has become more commercially oriented and enterprising in the last 25 years, large parts of it still remain subject to only limited, if not negligible, competition. Too often the instinct of the politician and administrator is to intervene, to try to manage, sometimes shackle, the market and to generally distrust its workings. A current example of this is to be seen in the proposal by the European Commission for regulating airport charges. This proposal, backed by airline trade associations, will if implemented have the effect of imposing unjustified economic regulation on many airports; in the UK, for example, a score of airports operating in a competitive environment will fall unnecessarily into its net. The alternative approach for moderating prices, by breaking up ownership groupings, encouraging entry and generally seeking to increase competition, largely is ignored.

Critical analysis of such policy initiatives is a feature of this book. It is based around a collection of my papers and other writings spanning the last 25 years. Subjects addressed include: the potency of potential competition; whether the experience with bi-laterals for intra-EU aviation prior to reform was likely to introduce more competition into thin markets after reform; the design of light-handed strategies for dealing with predation; whether some airports are natural monopolies and, if so, whether they will abuse their market power; whether economic regulation can replicate the outcome of competitive markets; whether incentive regulation typically applied to the utility industries, including some airports, leads to under or over-investment in infrastructure; whether high fixed costs force financial losses for small airports that justify public subsidy; and whether there should be *per se* rules governing airside access to airports.

Previously published material has been extensively edited and adapted and combined with new material published here for the first time. The book is divided into five sections, each featuring an original overview chapter, to better establish the background and also explain the chapters' wider significance including, wherever appropriate, their relevance to current policy issues. The papers have been selected to illustrate a general theme: the role of the market and its interplay with the development of economic policy in the context of a dynamic industry. The result, I believe, illustrates how market mechanisms, and particularly competition, can operate to successfully resolve policy issues.

David Starkie
Marsh Benham
December 2007

Acknowledgements

First and foremost I should like to thank the co-authors of papers that form the basis of four chapters in this book. Those who know Margaret Starrs will attest to the fact that she is a lady who knows her own mind, so perhaps my memory fails me but I can recollect only that the origins for Chapter 1 were based on a most amicable collaboration. As an economist in the South Australian Department of Transport, she was a tower of strength when defending our arguments in a sometimes heated debate in Australian policy circles. At this point, I should also like to thank Derek Scrafton, then Director General of Transport, South Australia. Derek was most generous in his support, not only of this research but also in other ways and I will always be grateful to him.

I first met David Thompson, my other co-author, at the Institute for Fiscal Studies (IFS) in London. The Conservative government's utility privatisation programme was then in full swing and analysis of its economic efficiency was on the research agenda at the Institute; privatisation of the British Airports Authority was a particular facet which fell to David and me to research. It was the beginning of a long professional collaboration which the joint authorship of Chapters 4, 7 and 8 illustrates. Working with David was a great pleasure and it has left me with a deep respect for his intellectual integrity. Also to be thanked is John Kay who, at the time, was Director of IFS. John brought his incisive intellect to bear on the airport privatisation issue and, although he left us to get on with the task, his influence was inspirational.

I would also like to take the opportunity to remember a young German economist, Martin Kunz, with whom I spent many a fruitful hour in discussion and whose influence is to be found particularly in Chapter 13. Martin died only days after one such meeting and it was a great honour for me to be asked to give the first of the memorial lectures in his name at the annual Hamburg Aviation Conference. Another series of memorial lectures are held to honour the name of Michael Beesley. Michael, a former tutor of mine, graciously invited me on two occasions to present papers at a series of London lectures he co-organised with Colin Robinson under the auspices of the London Business School and the Institute of Economic Affairs. These lectures form the basis of two of the chapters in the book. Sadly, Michael died before the second lecture was delivered but I regard Chapter 5 as a tribute to him.

Other acknowledgements relating to specific material in this book appear at the beginning of various chapters and included at the start of relevant chapters are thanks to the holders of the copyright of the original material, who have generously given their permission for re-publication. However, I would like to convey my special thanks to David Gillen, Stephen Littlechild and Ralph Turvey for all their encouragement and support over many years.

To turn the original ideas for the book into reality has required the invaluable assistance of a number of people. Judith Payne, in her role as Production Editor of *Fiscal Studies*, had already cast her professional eye over chapters published previously in that journal but I am also very grateful to her for having advised on style and grammar for the new material herein. The production team at Ashgate have also been a source of strength. Special thanks go to Guy Loft, Commissioning Editor, who unreservedly backed the original ideas for this book and took me most efficiently through the initial stages of the publication process. Pauline Beavers, Desk Editor at Ashgate, has also been a great pleasure to work with and I am very grateful to Pat FitzGerald for all her patience and understanding. Ashgate are also to be thanked for their flexibility in agreeing to cooperate with the IEA in the book's publication. In this context, my sincere thanks go to Philip Booth, Editorial and Programme Director at the IEA for his generous support. I also owe a special debt to Cento Veljanovski of Case Associates. Cento provided me with an important break in my early consulting career and has compounded his generosity by providing additional support for this book.

Finally, I would like to thank Ursula, my wife, who has given unstinting help with the different tasks of stitching together a story line spread across many documents, in different archives, written in various styles. I am truly grateful.

I alone remain responsible for all errors of omission, fact and interpretation.

PART I
Airline Competition

Overview

In 1949, the Australian federal government introduced what became known as the 'Two Airline Policy', initially by agreement but later endorsed by statute. This policy allowed only two airlines to operate trunk routes between the major cities and, for a time, it required the airline duopoly to operate identical equipment and to offer identical fares. In spite of some subsequent relaxation of constraints, by the 1980s the policy was much criticised for stifling competition and in 1987 the federal government announced its firm intention to remove restrictions on entry into inter-state aviation; it put the industry on three year's notice. This changing policy was encouraged by a number of factors including an agenda of general de-regulation in Australia during the 1980s (see Starkie 1989). But, de-regulation of the US domestic aviation industry in 1979 was a potent factor and was used as an example by those Commonwealth states pressing for a change in federal policy. South Australia was one such state and it had taken steps some years before to lead by example: it had de-regulated *intra*-state aviation.

Chapter 1 is based on a detailed examination of this de-regulated air service market within South Australia undertaken by myself and Margaret Starrs, then of the South Australian Department of Transport, and the results formed part of the South Australian government's case made in submissions to the federal authorities prior to the 1987 decision to de-regulate inter-state aviation. The chapter analyses what happened following de-regulation in South Australian with particular reference to the theory of contestable markets which was receiving much attention at that time. Many of the routes served sparsely populated remote areas with eight-seat Cessna 402s being the most common aircraft type. The analysis showed that new firms had entered the market, services had increased significantly, new routes had developed and, importantly, there was no apparent difference in the structure of fares on multi-firm and single firm (monopoly) routes. In the light of these findings we speculated that contestability theory might have been more robust than its critics had thus far acknowledged. Nevertheless, it remains an unusual if not unique study of competition in an air services market much of which was at the subregional (or 'third') level.

Chapter 2 turns attention to Europe and to the period immediately following the introduction, in 1993, of the 'Third Package' of measures liberalising European aviation. At that time, there was speculation on whether and how the liberalising measures would impact on the European scene and easy assumptions were being made based on events that had unfolded in the US in the (1980s) decade following de-regulation of the US domestic industry. The chapter takes a more considered view. It is a much abbreviated version of my 1993 contribution to the lecture series established at the beginning of the 1990s under the auspices of the Institute of Economic Affairs and London Business School by Professors Michael Beesley and Colin Robinson, and later known as the Beesley Lectures in honour of Michael

Beesley, a former tutor of mine who greatly influenced my thinking. This 1993 contribution (not published until 1994) is one of three that I made to the series and another will be found at Chapter 5.

The focus of the piece is the hypothesis that the bilateral air service agreements that governed services within Europe prior to liberalisation in 1993, had embedded parts of the European market with a potentially competitive structure. This difference was evident in statistics that had been compiled for mid-1989 comparing European aviation (prior to the 'Third Package') with the situation in the United States at that time, (which was about a decade after de-regulation of the US industry). The statistics showed that the proportion of 'thin' or less dense markets in Europe served by more than one carrier was greater than the corresponding proportion in the US, probably as a result of the bilateral system that required services by at least one airline established at each end of an international route. In the chapter it is argued that this structural difference, existing at the outset of European liberalisation, could lead to a more competitive structure in Europe and thus help to offset those differences between Europe and the US (such as shorter and fewer dense routes in the former) that had the effect of increasing average costs per kilometre in European aviation. The chapter concludes by criticising the residual controls on airfares, specifically those applying to on-demand fares unencumbered by restrictions on use. In particular, it points out that the cost benchmarks, used by the European Commission when appraising such fares, are inappropriate in an environment where airport capacity is constrained and demand for its use is often subject to quantity rationing.

The final chapter in Part I, *Chapter 3*, switches attention to the United States and the issue of predatory conduct in the US airline industry. In April 1998 the US Department of Transportation published proposed guidelines on predatory conduct in the domestic aviation industry. The Department had come to the conclusion that unfair exclusionary practices had been a key reason why new low fare carriers had not been able to gain significant entry into concentrated hubs to compete with incumbents in the US market. It was of the view that a not untypical response by the incumbent to entry was to flood the market with additional low fare capacity. To counter such behaviour by incumbent airline(s), the DoT proposed to examine conduct on a case-by-case basis and to institute proceedings when it considered conduct to have been unreasonable. The DoT invited comments on its proposal and the chapter is based on my response, which was published subsequently in the *European Competition Law Review*.

The DoT's regulatory proposal I thought intrusive and considered that the problem could be addressed in a more benign way. The essence of my counter-proposal was that, in the event of an entrant subsequently exiting the market, the post-entry capacity (but not the price(s)) of the incumbent's services was to be temporarily frozen or locked-in. The basic feature of the proposal was that it penalised only those incumbent airlines responding to entry by offering *excessive* capacity in relation to market demand; it continued to permit during the entry phase an unconstrained competitive response by the incumbent. Post exit, the incumbent would be free to fine-tune prices (fares) but the core price(s) would be driven by the extent of the capacity supplied to the market and it was this quantum that it was proposed should be temporarily frozen on exit of the entrant. This approach regulates with a very light touch because the competition authorities need only monitor post-exit capacity (aircraft seats) supplied to the market and this is a much easier task than monitoring prices (yields). This latter problem of information asymmetries was the Achilles heel of an earlier and superficially similar approach, unknown to me at the time, the essential idea of which was to commit the incumbent to a post-entry price (Baumol 1979). (A further weakness of the Baumol approach is that it does not prevent the incumbent reducing capacity to achieve a profitable outturn at the committed low level of fare(s) although, by doing so, the incumbent increases the scope for re-entry.)

In addition to the DoT, I also sent my submission to Professor Alfred Kahn, (the former Chairman of the US Civil Aeronautics Board) because of his particular interest in this subject and because I had previously worked with him in New Zealand on an aviation case. This led to extensive correspondence between us, wherein Professor Kahn tested my arguments at length. I was eventually able to convince him of the veracity of the approach proposed. This correspondence can be read in Chapter 15 of Forsyth et al. (2005).

Chapter 1

Contestability and Sustainability in Regional Airline Markets[1],[2]

Introduction

The Australian airline market is closely regulated. The 'two-airline' policy affecting trunk routes is well known, but less appreciated are state controls imposed on regional and local aviation. These controls exist in four of the six states; Victoria and South Australia are the exceptions. This distinction between the states was of little consequence prior to 1979. Until then the Commonwealth government also controlled regional air transport using powers under the Air Navigation Act of 1920. But in 1979, as a result of the Domestic Air Transport Policy Reviews, the Commonwealth decided not to hinder the development of competition in regional air transport (BTE 1981, p. 113). The result has been a much more competitive environment in both Victoria and South Australia.[3]

This chapter focuses on South Australia and considers what has happened since 1979 by reference to recent developments in the theory of market structures. These developments stress ease of market entry and exit and have particular relevance to industries with mobile assets such as aircraft. The South Australian regional airline market provides a good basis for testing theory. The overall size of the market is limited, there is a number of low demand or 'thin' routes and there is a considerable range of stage lengths. The thrust of the chapter is empirical and descriptive but [first] we set out the theory of contestable markets [and then] we analyze in broad terms the current South Australian position, before [...] focusing upon developments during the last four years. Finally, [...] we consider

---

1   First published (with M.M. Starrs) in *Economic Record*, 60 (170), 274–83; published here by kind permission of Wiley-Blackwell.

2   We would like to thank Ian Unsworth of the Independent Air Fares Committee; Barry Roberts and John Streeter, of the Department of Aviation; Derek Scrafton, Director General of Transport, South Australia; and Helen Wickens, Bronco Karagich, Kerry Clift, John Hatch and Ian McLean of the Department of Economics, University of Adelaide for help in compiling, processing and commenting on material. We are grateful also for the very useful comments of the editor and anonymous referees. We remain, of course, solely responsible for the views expressed herein.

3   Operators are still required to submit proposals to the Federal Department of Aviation for reasons of air safety. A benign approach also appears to have been adopted thus far by the Independent Air Fares Committee which, since late 1981, has had explicit powers of approval with respect to fares.

the general performance of the South Australian industry with reference to the theory of contestable markets.

Contestability: A Brief Review of Theory

The major contribution of the 'new' theory of contestability has been to show that the structure of an industry – the number of competing firms – may have no bearing on the degree to which production will be efficient and welfare maximised. Traditional theory has assumed otherwise. As Baumol (1982) points out, the received theory of market structures tends to view efficiency as a monotonically increasing function of the number of firms in an industry, with unregulated monopoly and perfect competition representing polar cases.[4] Conversely, the need to regulate to achieve an efficient price and output combination is seen by received theory to vary inversely with the number of firms; in an industry inclined towards pure monopoly (because of a limited size of market and substantial fixed costs leading to economies of scale), regulation is considered essential.

In contrast to this conventional viewpoint, the new theory of industrial structures revolves around the idea that the competitive pressures required for an efficient solution can come equally well from outside an industry. The key is an ability to contest for a market rather than to compete within it. This ability to contest depends upon the ease with which a firm can enter *and* exit from a market without cost which, in turn, depends particularly upon whether capital is mobile, irretrievably committed to producing a particular product. It is argued that the power of a firm to extract monopoly rents depends upon the extent to which production stems from immobile capital that is the extent to which the fixed costs of production are also 'sunk' costs. Sunk costs are costs facing a potential entrant which do not have to be paid *once more* by the incumbent; to the potential entrant they constitute a barrier to entry. On the other hand, if all capital is saleable and reusable in alternative markets without loss (other than that corresponding to normal depreciation in use) a potential entrant to an industry can then replicate, without penalty, the cost and output vectors of the incumbent firm(s). Consequently, an industry without sunk costs even a natural monopoly industry – is said to be a perfectly contestable industry; the possibility of entry by rival firms is a constant threat.

The ability to contest a market in these circumstances has a number of important consequences in terms of welfare. First, a contestable market, in long-run equilibrium, never offers more than a normal rate of profit. The existence of (temporary) supernormal profits will attract rival firms willing to offer the same output at lower prices. Consequently, a monopolist, providing his position in the market is perfectly contestable, will earn zero economic rent. Second, production inefficiencies also will be totally absent in long-run equilibrium – unnecessary cost (like abnormal returns) constitutes an invitation to entry. Third, in long-

---

4   Note also that X-efficiency has been shown to be a function of firm numbers (Hatch 1979).

run equilibrium, no product produced in a contestable market can be sold at a price less than its marginal cost. A price less than marginal cost will allow a rival firm to enter the market and offer a smaller output at a slightly lower price and yet, by eliminating the unprofitable marginal unit, earn at least as much as the incumbent. Consequently, in the long run, cross-subsidies and predatory pricing practices are unfeasible. And, fourth, if a market contains two or more firms, again, in the long run, prices cannot exceed marginal costs.

The only contrary note for maximising welfare arises in the case of a monopolist. It *may* be possible for a monopolist's price to exceed marginal cost if demand cuts the average cost curve in its falling segment, i.e., at an output less than qm in Figure 1.1 where for outputs less than q^* single-firm production minimises costs.[5] An attempt by an entrant to sell a greater ouptut at a lower price may be thwarted by a low elasticity of demand so that the market will not absorb the additional output at a price that makes entry profitable. Nevertheless, even though price may exceed marginal cost, a monopolist when strongly threatened by a potential entrant, will be inclined to adopt prices that reflect what the market will bear (i.e., Ramsey-optimal prices). Therefore, as Baumol remarks, a contestable market offers some presumption (but no guarantee) that a monopolist, required to cover total costs from revenues, will behave in a manner consistent with a second-best optimum, i.e., that inefficiencies will be minimised.

If it is not possible for the monopolist to price in excess of marginal cost, a natural monopoly that is contestable is no longer sustainable – i.e., there is no price-output vector such that entry by a rival firm is unattractive, all demand is satisfied *and* revenues cover total costs of production. This will occur if demand cuts the average cost curve in its rising segment, i.e., at an ouput between qm and q^* in Figure 1.1. It will be possible for a rival firm to enter the market and produce output qm at a lower cost. As a consequence either total costs of production will be raised (if total demand is to be satisfied by more than one firm) or consumer welfare will be reduced by restricting output to a suboptimal level.[6]

Apart from the natural monopoly case, contestable market theory appears capable of reproducing the efficiency of perfect competition without the same, demanding, pre-conditions. Perfect competition can be thought of as a special case of perfect contestability with the latter extending considerably the domain within which welfare is maximised. The ideas inherent in contestability are by no means unique. The affinity between contestable market theory and established literature on barriers to entry and workable competition, have been noted. But what appears to be novel is the stress placed on barriers to exit and the claim that

---

5   We are using Panzar and Willig's (1977) definition of a natural monopoly as a firm which is the sole seller of a good whose technology makes single-firm production cheapest given the size of the market.

6   In circumstances where the monopolist is also a multi-product firm, for example selling its output in different spatial or temporal markets, the issue becomes more complex. A multi-product monopolist may be able to sustain prices in all its submarkets or only in a few of them.

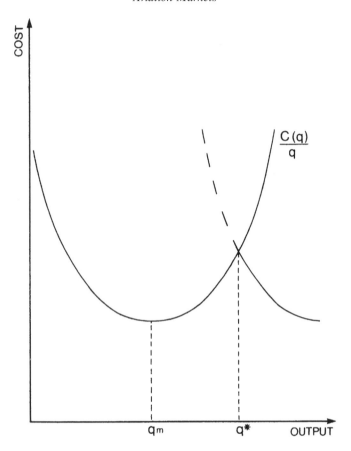

Figure 1.1   Single firm production, output and costs

the theory makes the structure of an industry dependent upon cost and demand
conditions and thus endogenous (Hatch 1983).

These claims are contentious and are subject to debate in the current literature.
Schwartz and Reynolds (1983), for example, argue that perfect contestability
assumes that the entrant can undercut the incumbent, capture the market and exit
before the incumbent can adjust his price. If this is not the case and price can be
adjusted instantaneously as rivals enter the market, the incumbent will set a pre-
entry price in excess of post-entry price. However, this, in turn, assumes that price
adjustment costs are zero (Starkie 1984). A further point of controversy concerns
the robustness of contestability theory and whether it breaks down completely
when sunk costs are present, even if small, i.e., do small sunk costs result in only
small deviations from optimal efficiency as Baumol et al. (1983) claim?

The contestable theorists' case in relation to the airline industry appears
to rest on three propositions. First, sunk costs are small and are not a serious
barrier to entry or exit. Second, if conditions in a particular airline market
favour a natural monopoly, then the monopoly (besides being contestable) will

be sustainable at prices that cover the firms' average costs and at output levels which fully satisfy the demand for air travel. And, third, the price adjustment lag faced by the incumbent exceeds the entry and exit lag of the potential entrant or price adjustment costs are positive. The first of these three propositions has received considerable support, the latter too little consideration. Bailey (1981), for example, argues that there is no reason, *a priori*, to expect economies of scale with respect to size of aircraft to lead to substantial barriers to entry because airline capital costs, while substantial, are not sunk costs. The chief sunk costs in aviation – runways, ground facilities and air navigational aids – are incurred as a rule by governments and not by airlines.[7]

Empirical analysis, thus far, has been equivocal. Using US data for 1979 and 1980, Bailey and Panzar (1981) showed that potential competition by trunk carriers appeared to provide an effective check on the prices of local service carriers operating long- and medium-haul routes. Graham et al. (1983), using more recent data for the most heavily travelled markets in the US, appear to arrive at an opposing conclusion. They argued that the degree of firm concentration (measured by the Herfindahl index) had a positive and significant effect on fares. However, their result was not unqualified. If concentration was modelled as an endogenous factor (in strict accordance with theory) the estimates became statistically insignificant.[8]

In the remaining sections of the chapter we examine the theory of contestability by reference to the South Australian airline market. Specifically, we consider the case advanced by a former executive of Ansett Transport Industries that the costs of developing routes could constitute an entry barrier (sunk cost) and that fares will differ according to the degree of competition in the market (Pascoe 1983). We also consider how the structure of the South Australian industry has changed since entry policy was revised in 1979 and how a route monopolist has reacted to entry.

Competition and the South Australian Market

In May 1983, ten South Australian based operators were operating scheduled services within the state.[9] These ten included Airlines of South Australia (ASA), an operating division of Ansett Transport Industries Limited. ASA uses F27s.

---

7   Nevertheless, governments can and do try to recover such costs from 'rental' fees.

8   We would add further qualifications. The authors noted in their final paragraph the important distinction between one and more than one firm per market. We think it more appropriate to have modelled this dichotomous effect rather than apply a continuous measure of concentration.

9   The commercial network excludes one route operated between Adelaide and station properties in the far northeast of the state and in south-west Queensland by Drennan Aviation. The route is subsidised by the state governments, federal government and Australia Post, and is not considered in the chapter.

The rest of the South Australian aircraft fleet is made up of different types of small aircraft with the eight-seat Cessna 402 being the most commonly used. The largest number of SA routes operated by one 'airline' is six (Commodore); ASA and O'Connor's four; and Rossair and Trans Regional Airlines three each. Five operators (Eyre Charter, Albatross, Emu, Opal and Skytours) fly a single route; in the case of Albatross and Skytours it is a simple end-to-end route with no intermediate stops.

The route structure is strongly focused upon Adelaide with all routes having Adelaide as the hub. In other respects the network defies easy generalisation. It is complex and subject to change. In the middle of 1983 there were 32 ports on the commercial network of scheduled services. Thirteen of these were served by more than one operator. However, the number of ports served by *competing* operators is ten – Kingscote by five, Broken Hill, Port Pirie and Port Lincoln by three and Cleve, Leigh Creek, Port Augusta, Whyalla, Renmark and Mildura each served by two.[10] A number of ports served by a single operator are within close driving distance of other ports and, in this respect, each may be considered to be serving a similar submarket. Kangaroo Island, for example, has four different ports on the network whilst the southern tip of the Eyre Peninsula has three. In the case of Kangaroo Island, consideration of submarkets does not alter the competitive balance (the total number of competitors is five, the same number that serve Kingscote) but, in the Eyre Peninsula case, it does have the effect of increasing the number of competing operators from three to four.[11]

Directly-competing schedules occur, for early weekday morning departures, for half the ten ports served by competing operators. Although there are competing schedules at other times, such occurrences are restricted largely to Kingscote; midday, evening and weekend schedules by an operator generally remain unopposed. In addition, competition on some early morning schedules is tempered by the indirectness of alternative flights.

In contrast to service competition, price competition is restricted. The constraint in this case derives from the *Independent Air Fares Committee Act, 1981*. The Independent Air Fares Committee (IAFC) is responsible for the determination of air fares on scheduled services (including intra-state services) by incorporated bodies. Since it was established, late in 1981, the committee has been preoccupied with fare determinations on the major trunk routes and its involvement in local markets has been of a more limited nature. Nevertheless, its chief concern with the small 'airlines' remains the same as that for the major operators – 'to ensure that the level of air fares is related as closely as practicable to the cost of providing the services for which those air fares are charged' (the *JAPC Act*, 1981, Part III).

---

10   Both Rossair and O'Connor's have agreements with ASA.

11   The close proximity of ports in these two submarkets nullifies the effect of both Cummins (Eyre Peninsula) and Penneshaw (Kangaroo Island) being privately owned ports with a potential to restrict entry to the market. In the case of Penneshaw the owner recently refused access to Trans Regional in favour of Commodore.

The *JAPC Act* allows one operator to match an approved fare of another operator on the same route without having to seek approval. If a potential entrant wishes to offer a lower fare than an incumbent, approval must be sought but, once accepted, the incumbent can, without delay, reduce his fare to match the new fare. Schwartz and Reynolds (1983) suggest that such a situation will lead to pre-entry prices exceeding post-entry prices and the incumbent reaping excess, but temporary, profits. Our view is that this will not necessarily occur if the price adjustment costs of the incumbent are positive.

Approval of fares by the IAFC involves examining fare applications and comparing the proposed fare with the distances involved and drawing attention to major departures from industry-wide standards. Naturally, a certain degree of 'reining-in' is to be expected, but the committee does allow also a degree of latitude when matching its own judgment against that of the operator. Consequently, it is possible to examine whether the contestability propositions are a reality in the South Australian market; specifically, whether there is a difference in the fare structure – albeit small in view of the IAFC's influence – between single-operator and multi-operator routes. If such a difference does prevail it would suggest that the market is not contestable – that either barriers to entry (or exit) exist or that the monopolist is encouraged by the absence of adjustment lags and costs, to post higher fares.

We examined this case by regressing the basic one-way economy fare against both great circle distances and a dummy variable. The dummy variable took the value 'one' if the fare was on a route sector flown by more than one operator, otherwise it was scored zero.[12] All fares, including those for travel between intermediate ports en route, were examined in this way. Just under half of the 62 fares included in the regression were for sectors flown by more than one operator; the mean distance was 305 kilometres (with a standard deviation of 211) and the mean fare $61.65 (with standard deviation of $42). The resulting equation was:

$$Y = 0.188 \, X + 4.135$$
$$(.008)$$
$$R2 = 0.9 \qquad n = 62$$

where: Y = fare ($), and $X,$ = distance (km).

The dummy variable was insignificant indicating no difference in the basic fare structure between single-operator routes and multi-operator routes.[13] The

---

12   The approach is similar to Primeaux (1977) except that he used estimated average firm-cost as the dependent variable.

13   A plot of the standardised residual against the predicted standardised dependent variable suggested that the relationship between fare and distances was linear as no transformations were tested. Aircraft type was not entered as a variable because we assumed that each firm would be operating at a load factor that offset the differences in the seat-kilometre costs of different aircraft. Examination of selected routes showed that light aircraft were achieving significantly higher average load factors than F27s (Starrs and Starkie, 1983).

'flag-fall' component (the constant term in the regression) was $4.14 and the distance rate ¢18.8 per kilometre with distance 'explaining' 90 per cent of the fare variation. Examination of residuals indicated that the two single-product firms (with one end-to-end route each) were little different from the average. Albatross had a small negative residual (below average fare) and Skytours' fare was about average for the distance involved. Both are in competition with other firms on their respective routes.

We also need to take into account promotional fares. These exist only on half a dozen routes. The important factor here is the extent to which promotions are an attempt to adjust fares in line with the lower marginal costs of some products. Apart from two stand-by fares, most concessions are offered by ASA. These are of two types, an advance purchase fare and a concession fare for the residents of Kingscote, Port Lincoln and Whyalla. The concession fare is not available from Ceduna and Mt Gambier, also served by ASA but in the face of limited competition. This would seem to negate the contestable market hypothesis. However, the fact that the concessions are available only in the off-peak winter season and are not available on ASA services flown by chartered Cessnas rather than the much larger F27s, suggests a genuine attempt to fill surplus capacity in the larger markets where product differentiation is feasible. An examination of the regression residuals also supports this view. The three ports with concessions have above average economy fares and the two ports without concessions below average economy fares.

Changes Since 1979

In this section we consider the degree of change that has taken place since the adoption in 1979 of a more liberal attitude to market entry at federal level. Table 1.1 compares the number of weekly flights and ports served on scheduled services by South Australian based operators in September 1979 and again in May 1983. There has been a net increase of three in the number of operators providing scheduled services; four firms having offered services for the first time since September 1979 and one, Williams Aviation, having passed into receivership.[14]

Three of the new entrants (Albatross, Eyre Charter and Skytours) are single-route operators; the exception is O'Connor's. During the last two years Commodore in particular has added significantly to their network and in the past few months has developed two routes totally outside South Australia (Mildura-Melbourne and Mildura-Broken Hill). Commodore's expansion has been the result of picking up some of the pieces after the demise of Williams Aviation and having taken over some of the assets and routes of the old Pagas company. Pagas' remaining assets and routes are operated now by Trans Regional Airlines.

---

14   Williams passed into receivership in February 1981. Informed opinion suggests that the demise of the company was a consequence of ownership passing interstate, the introduction of inexperienced management and a big expansion of operations incorporating up-market turbo-engined Metro IIs and Citation jets.

Table 1.1   South Australian scheduled services: 1979 and 1983

	September 1979		May 1983	
Operator	**Ports served**	**Weekly flights**	**Ports served**	**Weekly flights**
ASA	7	141	7	130
Commodore Aviation	2	24*	10	168
Emu Air Charter	3	56	3	28
Opal Air (SA)	5	41	5	26
Pagas/Trans Regional	8	98	7	52
Rossair	2	6	6	22
Williams Aviation	7	68	–	–
Albatross	–	–	2	24
Eyre Charter	–	–	3	20
O'Connor's	–	–	18	40
Skytours	–	–	2	12
TOTAL FLIGHTS	–	434	–	522

*   Average over the whole year.

Source: BTE (1980, Table 4.12) and analysis of current timetables.

Other operators, extant in September 1979, have on the whole maintained a fairly stable level of operation. ASA, for example, operate the same route structure although at a slightly lower frequency.[15] Emu and Opal, on the other hand, have contracted considerably their number of weekly flights although the number of ports served by each has remained constant. Taking this contraction of flights into account and the fact that the single-route new entrants have, thus far, shown little inclination to expand, suggests that, in spite of Commodore's behaviour to the contrary, economies of scope are limited. The co-existence of large and small 'commuter' operators – evident also on a national scale-appears to support this view.

Between September 1979 and May 1983, the total number of weekly flights into and out of Adelaide by 'commuter' operators expanded by one-third; the total number of scheduled flights increased by 20 per cent.[16] The data available

---

15   This constancy of ASA's operations is important because it allows for interpretation of changes in other services as net changes. It avoids, for example, the need to take account of quality trade-offs between non-pressurised aircraft and pressurised F27s.

16   The expansion of available seats was less; we ignore this aspect. The consumer's preference function will include fares, the number and timing of flights and in-flight quality and it is these aspects we take account of.

for September 1979[17] classifies ports by ranges of flights per week so that a precise comparison with the 1983 situation, port by port, is not possible. But a significant proportion of the increase in weekly flights appears to derive from the development of new routes to new ports of call rather than from competition *in* a particular market. Extra services also have been introduced on single-firm routes. Within the state, there are only three ports, including Mt Gambier, evidently worse off in May 1983 compared with four years previously. In the last few years road services between Mt Gambier and Adelaide have improved considerably and new air services eastwards into Victoria have developed also. These factors possibly explain why this port now has fewer weekly flights to and from Adelaide.

To consider how increased competition has affected flight frequencies, departure and arrival times and fares, we have examined in some detail developments in the comparatively large Port Lincoln-Adelaide submarket. In May 1983, there were four operators (ASA, Commodore, O'Connor's and Rossair) flying scheduled services between Adelaide and Port Lincoln. In addition, Eyre charter offered services from Cummins, about 50 kilometres north of Port Lincoln (with Tumby Bay as an optional port of call), while Trans Regional Airlines were about to commence twice-weekly flights to Port Lincoln (via Kingscote). However O'Connor's only picks up passengers at the request of ASA and Rossair is chartered by ASA to supplement its F27 services. Consequently, the number of operators competing *in* the market in May 1983 was three; in mid-1979 ASA was the sole operator.

Timetables for the route have been analyzed at eight points in time to obtain some indication of the changes in service levels that have occurred in the more competitive environment (Table 1.2). In September 1978, for example, prior to Commodore's entry to the market, ASA provided 40 services per week between Adelaide and Port Lincoln, 20 in each direction. Day-return trips to Adelaide by Port Lincoln residents were possible on three weekdays (Monday, Tuesday and Friday). Since September 1978, the number of flights per week has increased from 40 to 108 in March 1983, with a peak of 138, in February 1982. ASA have maintained their weekly flight frequency at 38 since 1980; this compares with 40 prior to the more competitive environment. The relative constancy of ASA's frequencies eliminates the need to interpret trade-offs between high quality F27 flights operating at a lower frequency and the increased frequency of lower quality commuter services.[18]

There appears to be little evidence of competitors scheduling services in parallel (Table 1.3). In September 1978 services were offered between 8.25a.m. and 6.10p.m., with an interval between flights of four and six hours.

By March 1983 services were offered between 7.30 a.m. and 7.10 p.m. with intervals in the late morning and early afternoon of 3.5 and 4 hours. There were

---

17   From Figure 4.4 in BTE (1980).

18   Forsyth and Hocking (1978) and Findlay (1983) amongst others have argued that there are increasing returns to users from enhanced service frequency (and that an efficient solution may require subsidies even if the airline industry is subject to constant returns to scale). We do not consider this issue but note that, since 1979, service frequency has increased on this and other routes.

Table 1.2   Flights per week between Port Lincoln/Cummins and Adelaide by operator, September 1978 – March 1983

Date	ASA*	Commodore	Eyre	Total
9/78	40	–		40
11/79	40	42		82
12/80	38	57		95
5/81	38 (4)	68		110
2/82	38 (14)	68	18	138
9/82	38 (8)	56	18	120
10/82	38 (8)	46	18	110
3/83	38 (6)	46	18	108

*   Figures in brackets show additional flights by Rossair's Cessnas on charter to ASA.

Source: Analysis of timetables.

Table 1.3   Port Lincoln/Cummins – Tuesday departure flights to Adelaide

					Date			
Time	9/78	11/79	12/80	5/81	2/82	9/82	10/82	3/83
0600–0659								
0700–0759		C	C	C	RC	C	C	C
0800–0859	A	A	A	A	AE	AE	AE	AE
0900–0959		C	C	C	C			
1000–1059								
1100–1159								
1200–1259	A	A	C	C	C	C	C	C
1300–1359								
1400–1459								
1500–1559						C		
1600–1659			A	AC	AC	A	A	A
1700–1759		C	C	C	C	C	C	C
1800–1859	A							
1900–1959		A	A	A	A	A	A	A
TOTAL	3	6	7	8	10	8	7	7

Notes: C=Commodore; A=ASA; R=Rossair; and E=Eyre Charter.

Source: Analysis of timetables.

attempts to reduce these intervals by scheduling additional flights but their subsequent withdrawal suggests insufficient demand.

Prior to Commodore's entry to the market, there was just the one standard economy fare offered on the Port Lincoln/Cummins-Adelaide route by ASA. Since Commodore's entry and Eyre Charter's commencement of services from near-by Cummins, the ASA economy fare has remained consistently the highest, increasing in real terms over this period. However, ASA has introduced a series of concessionary fares. Initially, the focus was on a discount for use of designated off-peak flights. Subsequently, an APEX fare was introduced (at a price similar to Eyre Charter's return fare) and, in the 1983 winter off-peak season, a Port Lincoln residents' concession fare was offered (matching a long established similar concession by Commodore). By May 1983, residents in the Southern Eyre Peninsula had six fares to choose from for a return trip to Adelaide.

ASA's reactions to the fares offered by the commuter operators must be seen against a background of falling passenger loadings and significant decline in ASA's share of the market since 1979. Passengers embarking on ASA's Port Lincoln services fell by 17 per cent although, during 1982, the position had stabilised and numbers increased slightly. During 1980 commuters held 15 per cent of the market, a share which increased to 31 per cent in 1982. ASA's experience on the Port Lincoln route is typical of their general experience in South Australia. Until 1979 they had operated for the most part as an uncontested monopolist and, therefore, they had most to lose from the advent of a contestable market.

The adjustment on the part of ASA has been interesting. ASA operate an all-F27 fleet in mixed 44-seat configuration and, consequently, they do not have an in-house flexibility to adjust the type of aircraft used on a particular service to anticipated loadings. Their reaction has been threefold. First, they have 'wet' chartered small Cessna aircraft from Rossair to supplement their own schedules. Second, they have shaved schedules operated by F27s. And third, they have transferred capacity to charter work, including a large contract ferrying workers in and out of the Moomba gas fields in the far north of the state. This entry and exit behaviour between markets has had the effect of maintaining load factors (Table 1.4) and maintaining if not improving slightly total revenue hours per aircraft even though a fourth F27 aircraft was added to the fleet.

Conclusions

Since 1979 there has been a significant increase in services offered in the South Australian air passenger market. New routes have been developed providing a number of centres with services for the first time while other centres, with services prior to 1979, have experienced an increase in frequencies. On multi-firm routes, emphasis has been placed on differentiation of the product by scheduling departure and arrival times to fill empty slots and, in the most competitive markets, by offering promotional fares. We do not know, of course, the extent to which these developments would have happened if more restrictive controls on entry to

Table 1.4   Performance indicators: Airlines of South Australia

	Revenue passenger load factor	Weight load factor (total)	Annual revenue hours per aircraft (RPT services)		Total hours on scheduled services
1982*	64.7	58.2	2332	1620	5703
1981	63.8	57.5	2306	1938	5989
1980	63.8	58.3	2129	2004	6333
1979	66.2	59.8	2178	2067	6200
1978	68.2	60.0	2315	**	**
1977	67.8	62.8	2114	**	**

*   1982 figures provisional.
**   Not available.

Source: Department of Transport Annual Reports and Department of Aviation.

the market had been maintained, but we are of the opinion that the new freedom of entry has acted as a catalyst.

About half the South Australian routes are single-firm routes, some new and some long established. There is little evidence in either case of hit and run entry having occurred and generally the monopolies appear sustainable. Multi-firm routes are mostly two/three firm routes with the major exception of Adelaide-Kingscote and Adelaide-Port Lincoln. These are subject to a more pronounced seasonal pattern of demand. Bailey and Panzar (1981) have argued that competition *in* the market will be more evident in tourist centres because the greater flexibility of the discretionary traveller, who is less concerned about schedules and the availability of a seat at the last minute, creates an opportunity to cover joint costs by offering a different product at a varied price. Load factors can then be made high enough to support the entry of several firms. The Adelaide-Kingscote/Port Lincoln markets appear to conform to this pattern.

Contestability places a premium on flexibility and to achieve this we would expect contestable markets to feature leasing (instead of outright purchase), chartering and subcontracting of specific services. This is apparent in the South Australia market with ASA's 'wet' charter of Rossair's Cessnas and Commodore's recent lease of a larger aircraft. This practice is assisted by what we discern to be limited economies of scope which allow, in the long run, for the co-existence of many firms of different size and thus a pool of operators available for subcontract. In this respect a close parallel can be drawn with the existing road freight industry.

In contestable markets when monopolists are strongly threatened with entry we would also expect to find fares on single-firm routes in-line with those on multi-firm routes. Analysis of standard fare rates on 62 route segments in South Australia (approximately half of which were operated by a single firm), showed that there was no significant difference between these two types of route in spite of there being no limitation on the speed at which incumbents could adjust prices. This suggests that price adjustment costs are positive, that sunk costs are very low or non-existent, or that contestability theory is more robust than its critics have so far acknowledged.

Finally, we can speculate on the applicability of our findings in the broader Australian context. The varied character of the South Australian market, the stage lengths involved, range of aircraft used and the size and nature of the submarkets – suggest that it provides a good test of conditions to be expected in other regional markets. The favourable consequences of minimal regulation probably would translate well to the other states with current restrictions on entry. The implications of our findings for the domestic trunk network are less obvious, given the different scale and type of operation. But we are encouraged to be more optimistic than recent commentators who fear that the Australian long-distance market is too thin to allow other than monopoly operations on many routes (e.g., Beazley 1983). Such an outcome is by no means self-evident. And, if it were to happen, recent developments in South Australia suggest that, in practice as well as in theory, thin markets, monopoly and efficient markets are not necessarily incompatible.[19]

---

19   Note that regulation carries its own welfare losses and that attempts to regulate monopoly profits can lead to cost padding (Albon and Kirby 1983).

Chapter 2

European Airline De-regulation: A Prediction[1]

The Framework

The regulation of European airlines is now relatively straightforward, at least in respect of intra-EC routes. Safety regulation continues much as it always has, but following the adoption of the Third Stage Package (Third Package) of European Council Regulations, the economic regulation applying to EC routes is arguably the most liberal in the world. From the beginning of this year (1984), airlines established in the EC and owned and controlled by EC nationals, require only a single Operating Licence, granted by a competent national authority, to fly on most domestic and European routes. Full access to member-states' domestic markets will be permitted by 1997. In addition, the distinction between scheduled and charter flights has been removed, and EC airlines are now free to charge whatever fares they wish on intra- EC flights without any need for prior approval (although there are regulatory safeguards regarding excessively high fares and fares which are predatory).

The CAA has been designated as the competent authority for granting Operating Licences in the UK. As with airport regulation, it has no statutory duty to promote competition. However, under the Civil Aviation Act 1982, it is charged with encouraging UK airlines to provide a range of services and with furthering the reasonable interests of airline users. And it is clear from various policy statements that the CAA regards competition within the framework of a multi-airline industry as the best way of achieving these objectives.

Against this background, the CAA can be expected to adopt a liberal stance in administering the new regime wherever it has discretion. For example, on the question of financial monitoring, its stated policy is to minimise intervention in order to concentrate on airlines whose financial failure would have the most impact. It has also introduced a separate Operating Licence for small-scale operators in order to minimise the regulatory burden on air taxis; before the new EC regulations came into force these services were exempt from licensing in the UK.

Of course, we are still a long way from 'open skies' the world over. In order to fly to destinations outside the EC, airlines also require separate Route Licences

---

1   Adapted from 'Regulating Airports and Airlines' in M.E. Beesley, *Regulating Utilities: The Way Forward*, Institute of Economic Affairs/London Business School, 1994. Published here by kind permission of the Institute of Economic Affairs.

and route access is still largely determined by bilateral negotiations between governments. However, the 'Third Package' is still a major step on the way to a more open and competitive airline industry.

The 'Third Package'

The major issue now is how well the 'Third Package' is working, what is to be expected from it, and whether the regulatory safeguards are appropriate. The regulatory safeguards concern mergers and acquisitions, the control of anti-competitive behaviour and monopoly pricing. I shall concentrate on the latter issues.

The view promoted by the popular press is that liberalisation in European aviation can be expected to lead to an era of cheap fares. Frequent reference is made to developments in the US where, following de-regulation, domestic fares tumbled not only in nominal but also in real terms as competitors flooded into the market. Of course, in recent years the US industry has seen a spate of mergers, acquisitions and insolvency, so that the levels of concentration in the industry are not all that different today than they were in, say, 1979. In spite of this, for both quality and price of the product, there remain significant differences from the regulated era. The picture is a complex one and prevents easy generalisation but real fares on the whole are well down on previous levels (although it must be remembered that at current levels of fares, collectively, US airlines are making losses). If comparisons are to be drawn I would suggest that it is this 'matured' stage of the US industry that forms a more realistic benchmark.

Against this benchmark one can note significant differences in the European environment: a much more constrained airport and airspace infrastructure; generally less dense and shorter routes; a culture of bilateralism, capacity sharing and revenue pooling; and an industry cost base which is high by US standards (although some of the inflated costs reflect the different operating environment). This suggests that perhaps we should not expect too much too quickly. It will take time for inefficient costs to be shaken-out (even where there is a will to do so), and in the short term the infrastructure constraints can only be eased. Therefore, I think it unrealistic to expect perfectly competitive behaviour (with notions of an equality between fares and marginal costs), especially where it is often the case that markets are thin, or where capacity is exogenously determined and limited by runway and airspace constraints.

Lessons from US Experience

What, therefore, can we expect? Here again, I think it is useful to turn to US experience. During the 1980s, there was a concentration of effort by US academics and others towards examining whether the airline industry was contestable – the idea that in the absence of sunk costs the possibility of hit-and-run entry produces perfectly competitive behaviour even in markets with few players. The

US industry was found not to conform to the contestability ideal; the number of airlines flying a route (the degree of concentration) did affect fare levels. More recently, analysts have turned their attention to examining competitive *conduct* on concentrated routes in order to explain pricing behaviour.

This is producing a rich vein of information, especially for duopoly routes which are characteristic of European aviation. What the studies have shown is that in the de-regulated US market where two airlines fly a route, the outcome as a rule is not perfectly competitive behaviour but neither is behaviour fully collusive; the tendency is for the airlines to be competitive to a degree. Put another way, we can say that on duopoly routes price competition is moderate rather than vigorous and that, importantly, collusion on price is absent. Other inferences can be drawn from this material. These include: the carrier with a higher share of a duopoly route market tends to price more competitively; carriers price more competitively on longer-distance routes and on routes that are predominantly leisure oriented; and, of course, the more competing carriers the more price competitive the route becomes.

Competition: Expectations in Europe

I believe that this type of information provides a useful point of reference when turning to expectations in Europe. It provides a guide for the regulator and it suggests that the regulator should focus on whether behaviour in the market is appropriately non-cooperative. The initial aim should be to undermine the culture that has traditionally prevailed in the European market and to get flag-carrying airlines competing in their duopoly markets. In this context I am reasonably optimistic that this can be achieved for reasons I shall now explain.

A recent analysis of the route structure of Western European schedule aviation by Richard Pryke of Liverpool University[2] has shown that, compared with the United States, European routes (including those in the European Economic Area countries) are less monopolistic where there are few flights. For example, half the European routes with 20–29 one-way flights per week have two or more carriers, whereas the corresponding proportion for the USA is only a sixth. This difference is explained by bilateral agreements which lead to both national carriers flying on international routes. But it does mean that, in effect, Europe has a potentially more competitive network structure; often thin routes are served by more than one airline. It probably also means that on a number of thin routes there are more seats offered than might be warranted by market demand. Couple this with the possibility that, in terms of productive efficiency, European airlines are currently *diverging* (with some but not all national carriers seriously seeking efficiencies), and one can see that there is a strong incentive for particular airlines to cast aside the traditions of cooperation and to seek advantage from a more competitive strategy.

---

2   Richard Pryke (1991).

This is merely a hypothesis and, of course, one can start from a different set of assumptions – that liberalisation will lead to more monopolies in thin markets (this was Pryke's working assumption) or that mergers and acquisitions will have the same effect. Much will depend upon how the regulator interprets his or her role and reads the signals correctly. On the latter there is perhaps some cause for concern.

Residual Price Controls

The Commission is in the process of putting in place safeguards against excessively high tariffs for on-demand fares unencumbered by restrictions on use. Unfortunately, the process appears to be developing a rather mechanistic approach. The principal instrument of analysis is the operating ratio defined as the relationship between total net receipts for a route and the costs incurred on the route. The Commission is inclined to presume an abuse of a dominant position when the operating ratio is exceeded by a specified amount. In addition, even if the overall ratio remains within acceptable limits the Commission may still judge that abuse exists if a 'basic fare' (the lowest fully flexible fare) is considered to be excessively high. To determine whether this is the case the Commission will compare this particular fare with all the direct and indirect costs attributable to the service while taking into account the return on capital as well as factors such as 'acceptable cross-subsidisation between routes'.

Personally, I am uneasy with this type of approach. The cost and revenue allocation problems are considerable, apart from which the process seems to contradict the intent of the liberalisation programme. It also appears to run counter to the dynamics of the competitive process; several studies have shown that in response to greater competition, an airline will actually increase the dispersion of its prices.

Moreover, it does raise a more fundamental issue. If airport capacity is constrained, as it is at many of Europe's principal airports, and it is not the practice for airports to charge market-clearing prices, is it not legitimate therefore for an airline to adopt price rationing? In these circumstances, fares may exceed direct and indirect costs by a substantial amount but the ensuing scarcity rents are not monopoly rents. In the face of these difficulties, it would seem preferable to focus *not* upon the level of fares or their variance, but upon the basic conduct of airlines operating various routes – more along the lines of the US analyses I referred to earlier.

Chapter 3

Predatory Conduct in the Airline Industry: A Proposal to the US DoT[1]

Introduction

In April 1998, the US Department of Transportation (DoT) issued a controversial proposal for an enforcement policy on predatory conduct in US domestic aviation.[2] Behind the proposal is a growing concern in the United States that the considerable reduction in real air fares achieved since de-regulation of the industry in the late 1970s was coming to an end. In the year to February 1997, for example, average fares rose 9 per cent,[3] and whereas previously decreases in discounted fares more than offset increases in unrestricted full fares, this was no longer the case. The US DoT's view is that this reversal of the previous trend in air fares is due, in large measure, to the difficulties new entrant carriers[4] face when competing with established carriers. New entrants face particular difficulties in penetrating the hub-based networks of the major carriers; the organisational structures of such networks make entry inherently difficult,[5] but the barriers are often reinforced by the 'exclusionary conduct' of the incumbents. It is the latter issue that is the focus of the DoT's proposal.

The Problem in Outline

Following de-regulation in the late 1970s the US domestic air carrier industry reorganised itself around a large number of hub-based networks.[6] Previously

---

1   The original was based on a submission made to the US DoT in response to a consultation process and was first published as 'The US Department of Transportation's Statement on Predatory Conduct in the Airline Industry: An Alternative Proposal' in *European Competition Law Review*, 20 (5), © D. Starkie.

2   Docket No. OST-98-3713: Statement of the DoT's Enforcement Policy Regarding Unfair Exclusionary Conduct in the Air Transport Industry.

3   American Express Travel Services.

4   Defined as independent airlines that have started jet services within the last ten years.

5   There are sunk costs in establishing the network, and entry on a scale necessary for effective competition is sometimes precluded by airport capacity constraints.

6   Although hubbing in the US predates de-regulation of the airline industry, after 1978 there was an acceleration of its use.

most journeys between US city-pairs were direct;[7] after de-regulation these were routed increasingly via an intermediate hub which became a point of (on-line) transfer between flights. The result was that most air passengers benefited from a more extensive, more frequent service and competition between air carriers over their respective hubs led to the inherited economies of a hub-and-spoke network being passed on to passengers in lower fares.[8] There was and continues to be, however, an important exception to the latter. Many hub airports are dominated by a single carrier, so that those passengers commencing their journeys at a hub pay significantly higher fares[9] than passengers commencing journeys at non-hub airports (from which alternative routings via competing hubs are often possible). Over time, hub dominance has grown; carriers have developed their hubs, a degree of consolidation between established air carriers has produced more single carrier hubs, and the major carriers have also grown adept at exploiting their local dominance.[10] The DoT noted in its proposal document, for example, that in local hub markets major carriers have focused on high fare services, leaving much of the demand for services at discounted fares unserved.

Such conditions should in principle attract competing carriers into the hubs to provide for the unserved discount fare market and to provide existing passengers paying fare premiums with more competitive offerings. These conditions should be particularly attractive to carriers with a low cost base, the archetypal example of which is Southwest Airlines. Southwest, originally a regional-based carrier, has grown substantially on the basis of a strategy that emphasises discount fares. It has been successful in developing a presence at an increasing number of hubs, whilst other new entrants have entered various local markets albeit on a more limited scale than Southwest. Overall, however, the impact of the new entrants has been limited, as recent increases in fare levels suggest.

The view of the DoT is that this limited impact is a consequence of predatory behaviour by the incumbent carriers.[11] The hub-dominating incumbent views competition by the new entrant carrier as a threat, in spite of which, in some instances, it chooses to co-exist with the low fare entrant and concentrate on developing traffic feeding into its network. On the other hand, a not untypical response by the incumbent is to seek to 'flood' the market with additional low-fare capacity. This strategy is often reinforced by the selective enhancement of

---

7   In 1978 domestic on-line connections accounted for 25 per cent and inter-line connections 23 per cent of total enplaned passengers.

8   Combining passengers with many different final destinations on the same flight into a hub enables the use of larger aircraft or, alternatively, higher load factors to be achieved (although there will be a fall in the average distance flown per flight sector).

9   After adjusting for length of flight and route traffic density.

10   A useful review is provided in P.P. Belobaba and J. Van Acker (1994).

11   The following summarises the DoT's stylisation of the pattern of events. There are contrary views which regard the stylisation as an inaccurate portrayal of the response to entry and challenge the basic concept of predation. Other commentators contend that this broad picture is based on a few isolated incidents, whilst others question whether the response of incumbents to entry has been the major reason for entrants' departure from local markets or for the failure of entrants in recent years.

frequent flyer miles for passengers and of commissions for travel agents. The result is that the entrant finds it difficult to establish break-even load factors and is thus forced to withdraw from the local market. After the entrant withdraws, the incumbent then drops the added capacity and raises fares. The major carrier thus accepts lower profits (or losses) in the short run, in order to secure a higher long-term return from its dominance of the hub. The strategy also benefits the major carriers prospectively: it signals to potential entrants that the major carrier will react aggressively to entry, and thus acts as a deterrent.

The DoT's Proposal

To counter such behaviour by incumbent carriers, the DoT proposes to examine conduct on a case-by-case basis (after receiving formal and informal complaints but also taking the initiative when appropriate.) It will consider whether a response to entry is consistent with behaviour in other competitive markets, especially those in which Southwest is a competitor.[12] In particular, when the incumbent sells a large number of seats at very low fares, or carries a large number of local passengers at the entrant's new low fares such that the entrant's total seat capacity, or passengers carried, is exceeded, the DoT will (unless there are strong arguments to the contrary) institute enforcement proceedings to determine whether the carrier has engaged in unfair exclusionary practices. In addition, the DoT will analyse other types of conduct, such as bonus frequent flyer awards or travel agent commission overrides, to see if incumbents appear to target entrants unfairly.

Notwithstanding this approach, the DoT does not intend to discourage major carriers from competing aggressively against new entrants in their hub markets. The Department suggests that a major carrier could, for example, match the new entrant's low fares on a restricted basis without significantly increasing capacity. It also comments that, conceivably, a major carrier could both lower its fares and add capacity without dilution of its existing revenue base[13] (presumably by price discrimination). And it is willing to concede that a new entrant's service might fail 'for legitimate competitive reasons'; in other words, withdrawal from the market by the entrant should not be construed to indicate predatory behaviour by the incumbent carrier.

It is clear that implementation of these proposals will require difficult judgments to be made. This is evident in the tension in the DoT's statement with regard to the addition of capacity by the incumbent carrier and in its reference to conceivable circumstances in which adding capacity in response to entry is a legitimate response. Indeed, the DoT's proposal recognises that 'a reasonable alternative' response by the incumbent could incorporate an element of revenue dilution, and

---

12   Southwest is used as a benchmark because its size and profitability are judged to make it a difficult target for exclusionary practices. Its presence on, or potential to enter, a route has been shown to impact significantly on yields (see K. Richards in B.S. McMullen (1996).

13   Referred to in the docket as 'self diversion of revenue'.

thus a great deal could hang on the interpretation of 'reasonable' conduct. The proposals also constitute a significant departure from established norms for judging predatory behaviour which have focused previously upon whether an incumbent, in responding to entry, has continued to cover costs. The DoT's proposal goes beyond these norms because an examination of whether the response by the incumbent has had the result of producing 'lower local revenues than would an alternative reasonable response' requires consideration not only of costs but also of revenues forgone. It is a recognition, therefore, that net revenues sacrificed by a course of action are an (opportunity) cost to the firm undertaking that action.[14]

The proposals have been severely criticised by the major airlines (as might be expected) and by a number of distinguished academic experts. The thrust of the more considered criticisms is that the proposals will have the effect of impeding the reactions of the incumbents to entry; where lower fares (following entry) stimulate an expansion of the market the incumbent will be inhibited from adding capacity and in the longer term the outcome could be higher rather than lower fares. These criticisms are arguable and have been robustly challenged. But there are those who, whilst strongly supporting what the DoT is attempting to accomplish, nevertheless doubt the 'administrability' of the proposals. In particular, it is questioned whether the major carriers can be expected to know when they are violating the measures which, if exceeded, lead to enforcement procedures. These measures require the incumbent to know the entrant's total seat capacity and the total number of passengers carried.

Back to Basics

What is required is a relatively simple rule that has the effect of making predatory behaviour unprofitable (or less profitable) without imposing significant economic costs on the industry. Let us start the quest by considering why conduct in the airline industry has failed to match the original ideas of contestability theory. When this theory was developed in the late 1970s and early 1980s, it was thought that the market for air travel was a textbook example of a contestable market, the essential idea of which is the ability to challenge for a market rather than necessarily compete within it. This ability to challenge or contest depends upon the ease with which a firm can enter *and* exit from a market without incurring a cost; which, in turn, depends particularly upon whether capital is mobile or sunk, that is, irretrievably committed to producing a particular product. Whether sunk costs exist in an industry is crucial to the argument. Sunk costs are costs facing a potential entrant which do not have to be paid *once more* by the incumbent;

---

14 See Professor Alfred Kahn Testimony: Kahn considers the DoT proposal superior to the Areeda-Turner test because it adds to the marginal production costs, upon which the Areeda-Turner is based, the opportunity costs of the incumbent's response. This is particularly pertinent when the incumbent adds capacity to the contested route because this capacity will, more than likely, have been directed from other markets and thus the incumbent will have sacrificed revenue elsewhere.

consequently, to the potential entrant they constitute a barrier to entry. On the other hand, if all capital is saleable and reusable in alternative markets without loss (other than that corresponding to normal depreciation in use) a potential entrant to an industry can then replicate, without penalty, the cost and output vectors of the incumbent firm(s). Consequently, an industry without sunk costs – even a natural monopoly industry – is said to be a perfectly contestable industry; the possibility of entry by rival firms is a constant threat, so that prices and levels of output are what one would expect to find under conditions of perfect competition.

These early claims became contentious and particular attention was paid by the critics to the speed at which an incumbent could respond to entry. A number of commentators argued that, to secure a competitive market outcome, it was also necessary to assume that the entrant could undercut the incumbent, capture the market and (if need be) exit before the incumbent could adjust its price.[15] If this was possible (and there were no sunk costs) then the threat of entry alone would encourage the incumbent to post competitive prices. However, the air carrier industry appears to be a good example of an industry in which sunk costs are low but incumbents are able to adjust prices almost instantaneously, and capacity almost as quickly, when entry occurs. As the DoT's Statement records: 'Compared to firms in other industries, a major carrier can price-discriminate to a much greater extent, adjust prices much faster, and shift resources between markets much more readily. ... Air carriers have access to comprehensive 'real time' information on their competitors' activities and can thus respond to competitors' initiatives more precisely and swiftly than firms in other industries.'[16] Airline capital might be 'capital on wings' but the machinery of adjustment is too well oiled; adjustments can be made too cheaply and too quickly for competitive threats to work effectively.

An Alternative Approach

But if it is here that the root of the problem lies, then perhaps the policy response should be designed accordingly. An alternative approach is to *lock in* the incumbent's response to entry so that if the entrant subsequently withdraws from a route the incumbent is required to continue to serve the route for a specified period without a reduction in the existing levels of service. During this specified period the incumbent would be required to operate with at least the same capacity (including frequency of service), as it did immediately prior to the withdrawal of the entrant.[17] The lock-in thus builds in a response lag. This has the effect

---

15   See, for example, W.A. Brock (1983).

16   The cost of adjusting prices (menu costs) is also negligible.

17   The reasons for focusing on capacity are threefold. First, to specify both capacity and prices (fares) is probably unnecessary; the capacity available in the market at a particular time largely determines the seats available at each fare. Second, the DoT's proposal remedies are essentially focused upon the problem of additional capacity. And third, it is easy to monitor and regulate capacity (this is essentially published information) but it is very difficult to do the same with yields.

of inducing the incumbent to post prices which one would expect to see in a competitive market. This is because if the incumbent's response to entry is to increase capacity to such an extent that losses are incurred, the incumbent would not be able to start recovering those losses immediately the entrant withdraws from the market. It thus prolongs the period over which an incumbent could incur losses, with the result that this form of predatory behaviour becomes more expensive to pursue and therefore is less likely to occur.[18]

The incentive compatible feature of the lock-in is that it imposes a potential but explicit penalty on predatory behaviour because it penalises *only* those incumbent carriers that react to entry by introducing levels of capacity which result in the carrier incurring losses in the short term. Any incumbent carrier operating a route which, post-entry, continues to make a positive contribution to the bottom line, will not be disadvantaged by the lock-in.[19] By, in effect, 'taxing' only predatory behaviour it increases the risks to the potential predator of adopting such behaviour. The proposed rule, therefore, does not eliminate predatory behaviour in response to entry but it is likely to reduce its occurrence. It enables the incumbent to continue to respond to entry by adjusting capacity but it reduces the likelihood that such adjustments will result in below cost operations which are unreasonable or unfair to the entrant.

The proposed rule requires a relatively light-handed form of regulatory intervention. The regulator is not required to make difficult judgments regarding the precise cost structure faced by the incumbent or (as the DoT's Statement requires) judge what would be 'a reasonable alternative response' by the incumbent in circumstances where major carriers are not to be discouraged from competing aggressively against new entrants in their hub markets. Instead, the regulator is required only to audit the incumbent's level of service on a route during the lock-in period. This should be relatively easy to achieve: the regulator can match the published OAG schedules with actual flight information.

The proposed rule is likely to be more effective in reducing predatory responses the longer the lock-in period chosen because then the greater will be the adverse consequences to those choosing to behave in a predatory fashion. In international aviation, schedules are usually set on a biannual basis and coordinated at scheduling conferences held twice a year. In this context, a six month lock-in period suggests itself. For US domestic aviation the same scheduling procedures do not apply. Schedule changes are more frequent and a three-month (90 day)

---

18   Professor Kahn has suggested that the lock-in does not capture a response which increases the availability of discounted seats on an unchanged total number of flights (and seat capacity). I feel that such a response is constrained in its effectiveness because the equipment used on a particular route is likely to have been optimised to the level of expected demand (prior to entry). Consequently, if the incumbent's response to entry is to increase the number of discounted seats on existing flights, this is probably at the expense of passengers wanting full fare unrestricted tickets. This might lead to a diversion of these passengers to the entrant.

19   Except in so far as [the carrier] would have chosen to seek super-normal profits during the lock-in period.

period might be more appropriate. However, the issue of an appropriate lock-in period merits further investigation taking into account seasonal factors in demand (which will affect certain markets more than others) and whether there are systematic changes during the course of a year in the price of factor inputs, such as aviation fuel.

Refinements

Seasonal changes both in demand and in the price of inputs purchased by carriers raise the issue of whether, particularly if a long lock-in period is chosen, there might be a release clause. The object of the lock-in is to reduce predatory behaviour by penalising the incumbent carrier who adds excess capacity and makes short-term losses in response to entry. In contrast, the incumbent carrier who responds to entry by increasing capacity and reducing fare yields only to levels expected in a competitive market (i.e. a profitable operation is sustained) would not be penalised by the lock-in, unless there were an adverse change in the underlying economic circumstances. If, for example, there were a reduction in market demand caused by exogenous factors, or a significant increase in the price of aviation fuel, one would expect air carriers to make corresponding adjustments to capacity (and prices). The lock-in would prevent such an adjustment and therefore could lead to possible losses even when the incumbent was pursuing a break-even strategy in response to entry.

There may be differing views regarding the 'fairness' of such an outcome. One argument is that such an eventuality should be treated as another business risk, which the incumbent carrier allows for in his post-entry strategy. There is, after all, the possibility that changing circumstances may lead to an outcome favourable to the incumbent: during the lock-in period demand may increase, or the price of factor inputs may fall. If the risks in this context can be viewed as symmetrical, (i.e. equal probabilities of demand/factor prices increasing/decreasing), then the presumption must be that the lock-in period does not penalise the incumbent who responds to entry by adopting a legitimate competitive response. If, on the other hand, these risks are not viewed as symmetrical, such that there is a greater likelihood of demand falling or the price of inputs increasing, there is an argument for allowing for this in the lock-in rule. For example, the incumbent could be permitted to alter capacity if it could show that it had been adversely affected by a change in external circumstances. However, a *de minimis* approach might be judged appropriate (thus acknowledging that the exogenous changes may also be favourable to the incumbent): only when adverse demand/cost impacts exceeded a specified threshold would the incumbent carrier be allowed to reduce capacity during the lock-in period. To simplify the administration, and to minimise delay in altering capacity in such instances, the incumbent could make changes without prior authorisation, but such changes would have to be filed and a selection of them subject to audit, with penalties for violation.

Summary and Conclusions

The US DoT's statement on predatory conduct in the airline industry is the latest attempt to address a problem which is concentrating the minds of regulators worldwide. Pinning down predation is recognised as a difficult task and very few actions against predation in the airline industry have been concluded in favour of the aggrieved party. The DoT's proposals are a focused attempt to provide guidelines on the issue and to specify criteria for the introduction of enforcement proceedings. Whether such enforcement proceedings would be more successful than in the past is questionable. The DoT recognises the fine line between a pro-competitive response to entry and one that is intended to drive the entrant from the market, and much will depend upon its judgment, especially regarding what is a 'reasonable' response in the circumstances of a particular case. The established airlines have strongly opposed the proposals and independent critics have questioned whether the criteria used to trigger enforcement proceedings are workable.

The DoT has focused its proposals on a particular response to entry by the incumbent carrier, that is, a response that floods the market with low fare capacity. With this in mind this article advances an alternative approach to the problem. It takes as its starting point a weakness evident in the application of contestability theory to the aviation industry, namely the absence of effective response lags which theory has shown are needed if prices and levels of output are to be kept at competitive levels regardless of the number of firms competing in the market. The proposal put forward in this article is to introduce a lock-in rule so that if the entrant does withdraw from a route, the incumbent is required to continue to serve the route for a specified period with pre-exit levels of capacity. It is argued that this approach imposes a potential but explicit penalty on predatory behaviour because it penalises only the incumbent carriers which have reacted to entry by choosing to operate with excess capacity. Incumbent carriers who react pro-competitively, possibly by increasing capacity but continuing to operate profitably, would not be penalised. A further advantage of the proposal is that it focuses the regulatory activities on monitoring and compliance; the regulator is not required to try to define in each case whether the response to entry has been predatory. The proposal appears, therefore, to meet the test of a relatively simple rule that has the effect of making predatory behaviour unprofitable (or less profitable) without imposing significant costs on the airline industry.

PART II
Airport Privatisation, Industry Structure and Regulation

Overview

At the time of writing, the co-ownership of three major London airports by BAA is subject to scrutiny by the UK competition authorities, a process that started in June 2006 when the UK Office of Fair Trading (OFT) announced that it was conducting a marketing study into the airports sector to see whether enhanced competition between airports would lead to greater benefits to consumers. The outcome of that initial study (OFT 2006) was the view that BAA's southeast airports could, under separate ownership, compete to attract air passengers and that a full inquiry into BAA's structure was justified. This inquiry is being undertaken by the Competition Commission and is expected to conclude some time in 2008.

Arguments concerning joint airport ownership were first exercised in the early 1980s at the time that the privatisation of the nationalised British Airports Authority was mooted. Many of those original arguments are set out in *Chapter 4*, which is based on a study report that formed part of a research programme on regulation and competition policy carried out by the Institute for Fiscal Studies (IFS) during the 1980s.[1]

The government of the day was not persuaded to break up the Airports Authority prior to its privatisation (although the Secretary of State, Nicholas Ridley, indicated in the House of Commons that the issue was finely balanced). The Nationalised Industry's Board and its powerful Chairman, Sir Norman Payne, lobbied hard for maintaining the *status quo*. This, too, was the proclivity of the Civil Aviation Authority (CAA) and of the Department of Transport, BAA's sponsoring department, partly for reasons that will emerge more fully in Chapter 5 and Part III of this book. The mid-1980s were, of course, early days in the privatisation programme of the Thatcher government; at the time the emphasis was on establishing shareholder capitalism through flotation of the major utility industries and there was also much faith in the ability of economic regulation to resolve problems of market dominance. By contrast, the countervailing view, arguing for restructuring of the industry prior to privatisation, was restricted largely to the IFS, the Adam Smith Institute,[2] consumer bodies and regional airports led by Manchester, who were fearful that their prospects would be harmed by a private airport juggernaut intent upon major expansion. Not surprisingly, the views of the incumbent management (the cooperation of which was essential for a successful privatisation), the Transport Department and the CAA, prevailed; the British Airport Authority became BAA plc in 1987.

---

1   The report was co-authored with David Thompson and published by the Institute in 1985 under the title 'Privatising London's Airports'.

2   Barrett (1984) was one of the first to argue a pro-competition case.

The utility of the chapter in the twenty-first century is, however, wider than its continuing relevance to the current UK competition inquiry. There are a large number of instances throughout the world where airports located close to each other are in common ownership, including major airport systems around New York, Paris and Berlin, but there are also many cases where a major airport is coupled with a lesser-used facility that might, nevertheless, be a potential competitor. One surprising characteristic of public policy regarding such airport groupings is the absence of a debate regarding restructuring to achieve a more competitive environment. Indeed, in Europe, proximate airports have been allowed to merge ownership with little such debate. With a slow, but sure, trend towards the transfer of airport equity to the private sector, this attitude that co-ownership and competition does not matter, may change. And, should it do so, the arguments in this chapter, particularly if the major airport systems are in focus, will be of wider relevance.

The chapter first discusses the salient features of BAA as it was prior to privatisation. It commends the organisation's franchising of commercial retail services at its major airports, suggesting that this leads to productive efficiency, but it is more critical of BAA's airside charges which it is suggested are set at a level below that required to remunerate investment at the margin. Its focus, however, is a scepticism that privatisation of BAA *en-bloc* would improve upon the *status quo* and it argues that the introduction of more competition between airports and between services at each airport, is an essential element in the case for privatisation. It also argues that introducing more competition would have only a limited (negative) impact on the proceeds that the government might expect to recoup from the sale of assets to the private sector. Nevertheless, if ownership was divided, there would remain an element of residual market power and, thus, a case for some form of economic regulation to be put in place. For the latter, the chapter considers rate-based regulation and the then new approach of incentive regulation operated through the application of a price cap; it favours the latter.

The chapter was co-authored over 20 years ago but, if I was writing it today (for example, as a submission to the current UK competition investigation), the fundamental argument would remain largely unchanged, especially with respect to airport competition. There are, however, two aspects where the emphasis might differ: a significant proportion of the argument is devoted to the lack of competition in retail services within the terminals and, consequently, a need to reduce or eliminate the resulting economic rent. Thinking on this point has moved on in two respects. First, it is arguable whether the economic rents from airports are any different in kind to those accruing in shopping centres such as Oxford Street, London; Fifth Avenue, New York; Rue du Faubourg St Hônoré, Paris and, thus, are quasi- rents that accrue to activities centrally located in relation to a spatial market. This argument is first developed in Chapter 5 wherein it is also suggested that there is an important complementarity between commercial revenues from airport retailing and airside charges, so that the existence of (location) rents from the former activity have a benign effect on the latter.[3]

The second aspect on which I would now take a different, less optimistic view is incentive regulation. The chapter comments that: 'The absolute or proportional size of the return on capital is not limited by this [RPI-X] method and thus the problem of over-capitalisation is avoided.' This, of course, was to adopt the pure incentive regulation viewpoint as originally espoused by Stephen Littlechild (1983).[4] Incentive regulation applied to the privatised public utilities has turned out to be different from that originally conceived by Littlechild; there is a strong aversion to utility industries earning profits in excess of the cost of capital so that: '[R]ate of return considerations are of central concern, albeit in a way which, in contrast to classical rate of return regulation, does provide some incentives to companies to innovate and cut costs' (Turvey 2001).[5] The difficult consequences that follow from the incorporation of the cost of capital into the analysis are traced out in a number of chapters in Part III, in particular in Chapter 11: 'Incentives for Airport Investment.'

The Conservative government published its White Paper on Airports Policy (Cmnd. 9542) setting out its proposals for Britain's airports in 1985. Reducing the role of the State was a key element and to effect this policy the government wanted to ensure that as many as possible of Britain's airports became private sector companies, including of course the British Airports Authority. But, airports were considered to enjoy 'local monopoly' (para. 10.12) and, because of this, it was proposed that major airport companies (whatever their ownership) should be 'designated' and thus have their air traffic charges regulated. The framework for this economic regulation of airports was established in the 1986 Airports Act, which, subsequently, set a benchmark for the transformation of the airport industry in other parts of the world including Ireland, Australia, Germany and South Africa.

Chapter 5 is a critique of the policies put in place by the 1986 Act. It was written in 1999 for presentation at one of the Beesley Lecture Series (see p. 2) and was thus able to take into account several price control reviews. Some of the material in the chapter had formed part of an Advice Note commissioned by the CAA to assist it with its early thinking for the 2002 Price Determination. Related material will also be found in Starkie (1994a) and (2001).

Initially it outlines the regulatory framework established by the 1986 Act, prior to discussing issues arising from this framework. The issues discussed are: the nature of the airport user in whose interest the regulatory system operates; the criteria used to specify which airports should be subject to price controls; whether service standards should be brought within the ambit of regulation; whether regulation should interest itself only in the level of charges disregarding

---

3   This argument has since found acceptance with the Australian Productivity Commission (2002, 371) and the UK OFT (2006, 5.20).

4   See also Bartle (ed.) (2003) for a number of commentaries on this issue.

5   Foster (1992) provides a good review of the earlier controversy regarding excessive profits of the newly privatised industries.

charging structures; and the problems caused by the adoption of the single-till approach (which takes into account the revenues from retailing, car parks etc. when setting the price cap).[6] The final section of the chapter discusses two reform options, first, a more focused form of regulation wherein the single-till approach is no longer mandatory for the price-capped airports and, second, the setting aside of price cap regulation.

In support of the reform options, it is argued that complementarities between airport retailing and property rents on the one hand and the number of airport passengers on the other, provides the airport with an incentive to expand air-side output. It is also argued that large airports are probably increasing cost industries, so that the regulation of prices based on normal or reasonable rates of return on capital might lead to inefficiently low prices. Both these points were subsequently echoed in the CAA's proposals for the 2003–08 Price Determinations: specifically, in the regulator's proposals for replacing the single-till with a dual-till approach and in proposals for a long-term price path. However, the review process leading up to the Determination was controversial and neither proposal formed part of the final Determination.

The controversial nature of the 2003–08 Determination, together with its main conclusions, is outlined in *Chapter 6*. Because of the polarisation of views during the price-cap review including a marked difference in the views of the two regulatory bodies, the CAA and the Competition Commission, the newly appointed Director of Economic Regulation at the CAA, Harry Bush, instigated a fundamentally different process in the lead-up to the 2008–13 price-cap review, a process now referred to as 'constructive engagement'. The chapter outlines the new process, is favourably disposed towards it, but foresees some major hurdles. The note of caution has since been borne out by subsequent events (with the low cost airlines in particular taking a less positive stance) but, on the whole, the new approach appears to have been well worthwhile.

Chapter 6 ends by raising again the issue first raised in my 1999 Beesley Lecture (Starkie 2000) of whether formal price regulation for airports is appropriate and, in particular, it argues that Stansted and Manchester might be treated differently from Heathrow and Gatwick. The suggestion has recently struck a chord with the House of Commons Transport Committee. In late 2006 the Committee recommended that the government review the continuing need for the designation of airports subject to economic regulation by the CAA as a matter of principle, and that it publish an assessment of the relative merits of this approach compared to the use of standard competition legislation to regulate the abuse of dominant position by airports.[7] It went on to suggest that the government should consider de-designating Manchester and Stansted as a first step (House of Commons

---

6   Incorporating commercial revenues alongside traffic revenues when assessing allowable revenues is generally referred to as the single- or common-till approach; excluding such revenues is referred to as the dual-till approach.

7   This recommendation followed closely my evidence to the Committee, see paras 124–5 and 128

2006).[8] The government responded quickly: the following month it announced that it proposed to consult on the issue, a process that started in February, 2007. It is expected that, after further consultations, a decision will be made in 2008.

---

8   The Office of Fair Trading in its Report on *UK Airports*, published in December 2006, also recommended that the government review the benefits of price regulation at Manchester.

Chapter 4

Privatisation and Structure[1]

Introduction

The present [1985] pattern of ownership of civil airfields in Britain is a complex mixture reflecting in part the ebb and flow of postwar policies towards public ownership. The 1983 Conservative Manifesto pledged that 'as many as possible of Britain's airports shall become private sector companies'. Airports are at present owned by several different public sector bodies: the British Airports Authority (BAA), local authorities, and the Civil Aviation Authority, and there is one significant privately owned airport (Southampton). However, BAA's London airports (Heathrow, Gatwick, and Stansted) handle nearly two-thirds of UK passenger traffic and issues regarding the effective use of capacity and investment are most sharply identified in the southeast.

The activities of the London airports can be divided into two broad categories: first, and most obviously, the airports provide for the landing and handling of aircraft and passengers; second, they provide a range of commercial services for passengers (such as banking, car-hire facilities, and duty-free shops) and for airlines (e.g. refuelling). These two groups of activities are of equal importance in the total turnover of London's airports. However, BAA does not itself provide many commercial services. Instead it franchises these operations to specialist firms in the private sector (for example Trusthouse Forte); BAA's income from these services includes fees for the franchises and payments for services such as heating, electricity, water, etc. Thus the total turnover (or customers' expenditure) on the commercial services at major airports is many times greater (seven times greater at Heathrow) than the turnover earned from landing fees and from handling aircraft and passengers.

Because many activities are not carried out directly by BAA but are subcontracted to specialist companies, BAA is different in character from other nationalised industries. Only 10 per cent of the workforce at BAA's airports [comprise] direct employees of BAA; the remaining 90 per cent work for the airlines, for franchisees, and for companies providing contracted services. The main tasks undertaken by BAA's direct workforce relate to airport security,

---

1 This chapter summarises findings in David Starkie and David Thompson, 'Privatising London's Airports', 1985, IFS Report Series 16. The chapter itself is adapted from Chapter 11 in John Kay, Colin Meyer and David Thompson (eds), *Privatisation and Regulation – the UK Experience*, Clarendon Press, Oxford, 1986. The Appendix to Chapter 4 is taken from the report. Both are published by kind permission of the publishers and IFS.

portering services, and the provision of emergency services. Many franchised activities are subcontracted to a particular company for a limited duration on the basis of competitive tenders; at the end of the period the 'franchise' is re-opened to tender. In other cases (for example aircraft refuelling at Heathrow), services are provided by a range of specialist companies acting in competition with one another. Both approaches might be expected to keep unit costs to a minimum and it can be concluded that, at least in relation to its commercial services, BAA is achieving productive efficiency (although these incentives do not apply to the infrastructure which BAA provides for its franchisees).

Its achievement of allocative efficiency is more questionable. BAA's policy for air traffic services is to set charges on the basis of long-run marginal costs. Because Heathrow in particular is operating at capacity for long periods, a more efficient solution would be to set prices to ration scarce capacity. However, analysis shows that BAA's stated policy of charging at long-run marginal cost does not appear to have been applied consistently or thoroughly. In 1980 BAA carried out a detailed study of the incremental costs incurred at both Heathrow and Gatwick Airports in preparation for legal proceedings brought against BAA by a number of airlines who argued that BAA's charges were discriminatory. The results from this analysis can be used to make a comparison between BAA's actual charges and the level of charges required to cover long-run marginal costs. Table 4.1 shows the revenues which it is estimated would be generated by setting charges equal to long-run marginal costs and compares these with the revenues which would be generated by the actual charges prevailing during 1983/4. The calculations indicate that charges in 1983/4 at Heathrow and Gatwick were substantially below BAA's own estimate of its long-run marginal costs in 1983/4.

BAA's objective in the case of its commercial services is to maximise profits (qualified by its public enterprise obligations and the long-term credibility of its pricing policies). We have noted already that BAA promotes productive efficiency in the operation of these services by franchising them to specialist companies in the private sector on the basis of competitive tenders. However, to achieve its aim of maximising the profits earned from these services BAA does have an incentive to reduce or minimise the degree of competition that the franchisee faces. With reduced competition, franchisees will be able to charge higher prices than they would otherwise be able to do and, consequently, they are able to increase their bids for the right to operate franchises. In practice the degree to which BAA restricts competition varies. The services provided to airlines at London's airports (such as refuelling) are moderately competitive. In contrast, most services to passengers (such as retailing) are subject to little competition within a terminal, and face only limited competition from off-airport or in-flight facilities.

To summarise, BAA appears to have chosen to set charges for air-traffic services well below marginal costs. In the case of many commercial services, competition has been restricted to ensure that prices are raised well above costs. Both policies are inefficient in allocative terms and have the effect that scarce capacity is not allocated to the most beneficial uses. Also, under-charging is exaggerating the requirement for costly new investment.

Table 4.1 **Traffic revenues implied by cost analysis compared with 1983/4 charges (1982 traffic pattern) for Heathrow and Gatwick Airports (£ million)**

Cost/charge category	Heathrow		Gatwick	
	April 1983 charges	**Cost-based charges**	**April 1983 charges**	**Cost-based charges**
Aircraft weight and movement	20	10	6	6
Passenger charge (inc security)	52	86	17	29
Parking	20	34	4	15
TOTAL CHARGES	92	130	27	50

Source: BAA 1983b, Table 10.1; BAA 1983a, Table 8.1.

Competition and Privatisation

There are a number of factors which might have contributed to this state of affairs. First, as a public enterprise BAA is not expected to maximise profits and is sheltered from take-over by organisations identifying opportunities for increased returns on capital. The government does set a financial target (the agreed target for the London airports for the current period is a minimum rate of return on average net assets of 3 per cent plus a growth-related increment), but it is difficult for the government to judge how efficient the Authority could be. Although the Authority will aim to achieve the agreed targets, it has no real incentive to maximise its return and therefore cover costs across all its outputs.

Second, the Authority is subject to pressures from a range of organisations which seek to modify its practices in their favour and the statutory consultation requirements placed on the Authority provide a convenient avenue for such pressures. The airlines in particular are adept at putting pressure on the Authority. They gain substantially from the Authority failing to charge economic prices for the use of traffic facilities and consequently they are prepared to expend substantial resources on keeping down traffic charges. Because the Authority, unlike many airlines, does not itself seek to maximise profits and is not subject to pressures from commercial investors, its ability and its resolve to resist pressures to act inefficiently are heavily compromised.

Third, and related to the preceding factors, because the Authority is not required, or motivated, to maximise the return on its invested capital, it could have an incentive to pursue the alternative goals of maximising output or maximising the scale of its capital assets. Its under-charging of traffic services (cross-subsidised from retailing activities) and its powerful advocacy of an enormous investment in Stansted in the absence of clear signals by the market that such an expansion is justified financially are features consistent with such goals.

Privatising BAA in its present form would not, however, be an adequate solution: the London airport system has significant market power and privatisation as a system would suggest a need for regulation on a substantial scale. But experience in the US shows that tightly regulated private monopolies display many of the weaknesses which we have identified in BAA. In addition, the threat of take-over to such a large and specialised organisation is unlikely to be strong. There is a danger that a privatised BAA would retrench on the policy of franchising commercial services (although a strictly profit-maximising management is likely to retain large-scale franchising). For these reasons the efficiency incentives resulting from *en bloc* privatisation are likely to be weak.

A more promising option is to privatise in a manner which promotes competition. There is considerable scope for increasing competition in the southeast provided that the ownership of BAA's airports is divided. Separate ownership of BAA's London airports would introduce more competition into the large market associated with travel to and from London and the southeast region. Stansted and Gatwick have the potential to compete strongly with each other (and with Luton) in the large and 'footloose' inclusive-tour and intercontinental discount fare markets where cross-elasticities are high. In addition, Gatwick is shaping up as a promising competitor to Heathrow for scheduled traffic; it now serves more UK regional centres and more cities in the US than does Heathrow and if the proposed services from … [London City Airport] … obtain approval, this will add a further, albeit small, competitive element.

Separate ownership will not eliminate monopoly power at the southeastern airports. Heathrow especially will retain such power to a considerable degree in specific segments of the market. But, with divided ownership, these segments will be fewer and, overall, the demand curve for anyone airport will be much more elastic than if Heathrow, Stansted, and Gatwick are maintained in unified ownership [see Appendix, p. 48].

BAA has argued that breaking up the Authority's airports will introduce a number of disadvantages, but these seem to be more apparent than real. There do not appear to be any marked economies to be gained by keeping the airports as one management unit. Nor do we foresee, from an investment perspective, problems arising. It is possible to expand many airport facilities by small increments and we would expect this to become a more common approach under a competitive regime. On the whole, the investment problems alluded to are not, in either content or scale, really different from those faced by a number of other economic sectors in which large sunk costs and competition are the norm, e.g. large-scale retailing, and chemical industries.

However, competition between airports is most unlikely to reduce the monopoly rents currently earned from commercial services. At Heathrow and at Gatwick, where there are multiple terminals, it has been suggested that these assets might be set up on a competing basis. If passenger services within these terminals continue to be awarded to franchisees on an exclusive basis this is unlikely to be effective. A better option would be to introduce or extend competition in the actual provision of these services within the terminal building, although in some cases there are physical constraints which make this difficult to achieve.

At the present time a small degree of competition is introduced by off-site facilities (e.g. in-flight duty-free sales by airlines and car-hire firms operating from perimeter hotels). A more radical departure along these lines would be to permit the sale of duty-free items at selected retail outlets in city centres. This is done on a large scale in Australia where it has had the effect of reducing prices across a wide range of tax-free goods.

There is a strong case, however, for regarding the duty-free concession as a tax distortion. The rationale for this tax concession is unclear; it is unlikely that it was intended as an implicit subsidy for the airport industry. The offer price for the airports could be expected to reflect these potential rents thus redirecting them to the Exchequer. But it is preferable that the government awards (directly or indirectly through an appointed agent) the trading concessions for duty/tax-free goods so that fees from the concession flow directly to the Exchequer.

Overall, we see the introduction of more competition between airports and within airport services as an essential element in the case for privatisation. Privatisation without establishing a more competitive structure will require a degree of regulation and intervention that must place a question mark over the basic case for privatisation. Thus, if dividing the ownership of the London airports is ruled out, to retain BAA's airports in the public sector and use the various powers the government has to encourage a more efficient outcome is an option which must be given serious consideration. Such intervention would be unpopular, it would be contrary to the present government's predilection to intervene less, and it would place the government increasingly in the role of second-guessing the market. But if it is not possible to establish a competitive market this might be the best way to proceed.

Regulation and Privatisation

Dividing the ownership of London's airports would introduce more competition but individual airports would still retain a degree of market power; there remains a case for adding to existing regulation. For example, it has been suggested that there should be rate-of-return regulation exercised by a specialist body similar to the Office of Telecommunications. The basic idea of rate-of-return regulation (a common practice in the US for regulating public utilities) is that the regulated firm is allowed a 'fair' rate of return on its 'rate base' – a measure of the firm's capital assets. A major limitation of this approach is the well-known over-capitalisation or Averch-Johnson (1962) effect. The firm can increase total profits by expanding the assets (rate base) on which a proportionate return is allowed (provided that the allowed rate of return exceeds the cost of capital to the firm). Consequently, there is a tendency towards over-investment.

Sherman and Visscher (1982) have shown that rate-of-return regulation also has an effect on the structure of prices – it encourages the use of multipart tariffs, price discrimination and, in some circumstances, the setting of marginal price below marginal cost for those activities which require marginally more capital. Therefore, rate base regulation results in inefficiencies of the type that appear to be associated with the current system of specifying for BAA a target rate of return on assets. Indeed, if there are pressures not to exceed the target or penalties in doing so (e.g. the threat of litigation from disgruntled airlines), then the target rate of return will produce exactly the same undesirable effects as rate-base regulation.

The alternative to rate-base regulation that has been suggested is to regulate monopoly power by specifying an acceptable rate of growth in charges and tariffs (Littlechild 1983). This approach has been adopted in the case of British Telecom and articulated in the form of 'RPI minus X' (i.e. increases are permitted only up to a level set at X percentage points below the change in the retail price index). The absolute or proportional size of the return on capital is not limited by this method and thus the problem of over-capitalisation is avoided. There remains an incentive to efficiency.

Applying this approach to BAA's airports suggests that prices charged by airport retailers could be subject to an RPI-X formula; charges for traffic services (which are set below marginal costs) could be subject to a formula once prices had risen to cover marginal costs.

The regulatory question has an obvious implication for the monies that the government will realise from the sale of BAA's assets. Restructuring BAA's assets prior to privatisation also raises questions as to whether and to what extent there will be an effect on the aggregate sale proceeds received by the Treasury.

Increasing competition in the market reduces the potential scope to generate monopoly rents. For this reason it may be argued that the options which enhance competitive forces in the airport industry will have an adverse effect on the sale price of the assets. A desire by the government to maximise the proceeds from privatisation appears to be in conflict with the objective of increased efficiency; if the southeastern airports are divided between different owners and this increases competition between, for example, Gatwick and Stansted, then privatisation proceeds will be reduced. This is to oversimplify the issue. If the profits generated by a privatised BAA are regulated (either directly or through the regulation of prices) then dividing ownership and enhancing competition will have only a limited impact on the sale price.

There also appears little substance to the notion that, because divided ownership reduces the opportunities for economies of scale or cross-subsidisation, it also reduces the sale proceeds. First, it is doubtful that there are significant economies of scale to be realised by multiple airport ownership. Management and administrative overheads, in which some savings from common ownership might appear feasible, account for only 10 per cent of total expenditure. In addition, our analysis indicated constant returns to airport numbers in this category of expenditure. Second, it makes little difference whether loss-making activities are wrapped up in a large asset portfolio or disposed of separately. Because losses can be offset against profits for tax purposes a broad portfolio will reduce the tax burden and thereby increase the sale price. A narrow portfolio (i.e. Heathrow, Gatwick, and Stansted separated) will increase the Treasury's tax receipts but will reduce the sale proceeds. The pluses and minuses will tend to balance out to produce a similar sum (provided that similar discount rates are adopted in the public and private sectors).

Existing capacity at Heathrow and Gatwick is limited relative to peak demands. Because of this, airside access to Heathrow and Gatwick is valuable and should be allocated by price to those who value access most. A market in access should be established. This can be done by allowing incumbent airlines to trade (via a broker to reduce the risk of predatory behaviour) their existing rights, although assigning resale rights to the airlines will affect the market value of the airport assets.

A final, and important, factor to take account of is the possibility of litigation – a repeat of the Pan Am case. This can be expected to have a depressing effect on asset values. The previous dispute was 'set aside' and a future action could, it seems, invoke previous contentions. The dispute was set aside on the basis of a Memorandum of Understanding between the American and UK governments. None of the preferred privatisation policies we have discussed appear to contradict, in principle, the elements of the Memorandum. This does not prevent a different interpretation by the airlines and, therefore, the threat of litigious action can be expected to remain in the background. There is no simple solution to this problem, although one way forward would be for the UK government to adopt the risks involved. It could do this by assuming liability for past claims set aside and it could agree also to accept responsibility for a proportion of any settlements that might arise from new claims.

Conclusions

Improving the efficiency with which resources are used in the airports sector depends critically upon increasing the degree of competition between individual airports and, in the case of commercial services, at each airport site. A privatised, competitively structured airport industry will be less inclined to develop facilities without proven demand and will be less inclined to cross-subsidise and view size as an end in itself.

One can speculate endlessly on what might have been, but we suspect that if London's airports had been privatised and their ownership divided two decades ago, the Stansted saga would never have occurred and certainly not in the form in which it has. Even now, it is unlikely that an independent, privatised Stansted will push to increase capacity to 15 million units or even half that figure. It is not evident that existing forecasts of demand will be sustainable if Stansted's users have to cover the costs of expanding capacity. And it is not certain how quickly demand will grow if charges at Heathrow and Gatwick too are based 'on sound economic principles' (Memorandum of Understanding 1983). At the moment, these charges are far from being economic or soundly based.

Appendix

Competition and Divided Ownership

[...] BAA faces an inelastic demand in some of its sub-markets and especially in the market for journeys starting or finishing in the London and southeast region.

As a profit-maximising enterprise, BAA would have an incentive to increase prices and restrict output in these markets. This is illustrated in Figure [4.1], the top half of which represents an airport with spare capacity (e.g. Stansted, Gatwick off-peak) and the bottom half a comparatively congested airport such Heathrow. Curves DA and DB represent demands under unified ownership. These demands will be relatively inelastic and of the same elasticity. They each represent fractional parts of the same market demand and with common ownership of the three airports it is assumed price changes would be coordinated and concurrent. The overall elasticity of the D curves is determined by competition at the margin (e.g. from Luton, Birmingham, Schipol, etc.). BAA will maximise profits by equating marginal cost and marginal revenue, the latter shown by the dashed curves, and by charging prices Pi^A and Pi^B. In both cases output is restricted below its welfare-maximising level. Shifts in demand will lead to changes of output, leaving prices unchanged (assuming constant costs up to capacity).

Dividing the ownership of the three airports between non-colluding owners will result in an increase in demand elasticities. The curves dA and dB show the demand/price relationship at each airport on the basis that each owner acts unilaterally and assumes that competitors will sustain prices. The increase in elasticities results in profit-maximising prices Pe^A and Pe^B. These are lower than the profit-maximising prices under unified ownership. Note also that at the relatively congested airport there is an actual increase in output.

The illustrated case is a specific version of Chamberlin's (1962, 81–100) analysis of group equilibrium in monopolistically competitive markets with product differentiation.

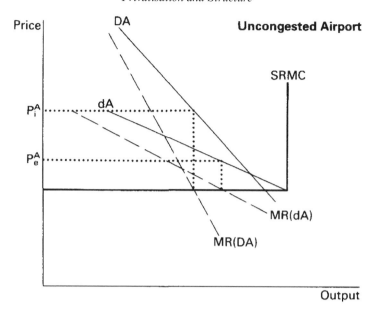

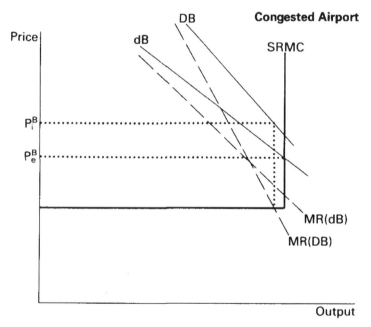

Figure 4.1   Airport pricing and competition

Chapter 5

Reforming Airport Regulation[1]

Introduction

Airports are different from the other regulated utilities, or so it would seem. The debate on the fundamentals of the UK approach to regulating its privatised utilities has focused on electricity, gas, telecommunications and water, but airports tend to receive a passing mention, if that. This low profile has particularly suited BAA and, when it has run the risk of being pigeon-holed with the other utilities, it has itself argued that it is different. It did so, for example, when the windfall tax was on the cards. It was not successful, however, in ducking the tax and it also received a shock at the time of this year's budget [1999] when the Chancellor announced a special review of the competition issues surrounding BAA's ownership of most of London's airports.

But there is some justification to the claim that airports are different and it was these inherent differences that led to Michael Beesley, in last year's series, suggesting a very different approach to regulating airports. This evening I want to continue with this theme. I, too, want to point to some important differences that set airports aside from the other utilities and I will argue that, to a degree, this reflects some curious or less common economic characteristics of airports. This will lead me to propose significant changes to the present regulatory regime and, perhaps, *in extremis*, to proposals not too different from those put forward by Michael Beesley. But, before doing so, let me sketch out the regulatory approach to airports. This too is different from the approach adopted for the other utilities and associated with it are a number of issues which I will outline.

An Outline of Economic Regulation

General Powers

The economic regulation of UK airports is governed by the 1986 Airports Act, but this legislation also includes more general powers that have the potential to

---

1   This chapter is based on joint work with Professor George Yarrow to whom I am indebted. I am also grateful to David Thompson for his comments. All views expressed here, however, are entirely my own. The paper was first presented at the 1999 series of Beesley Lectures and published as 'A New Deal for Airports?' in Robinson, C. (ed.) *Regulating Utilities: New Issues, New Solutions*, Edward Elgar; it is republished by kind permission of the Institute of Economic Affairs.

impact significantly upon the business of the airport companies. For example, the government is able to direct different types of air traffic to different airports within a 'single system' and, using these powers, until the early 1990s, aircraft carrying only charter passengers were not permitted to use Heathrow Airport. Other provisions in the Act enable limits to be placed on the number of air transport movements at congested airports. This might seem uncontroversial, but airports long since considered congested have shown a remarkable capacity to handle, over time, increasing numbers of aircraft (assisted by improvements in techniques of capacity management). The government, therefore, has powers under the 1986 Act (should it so wish to use them) to restrict both the volume and type of business handled by an airport company. Arguably, these powers allow for greater intervention in the airport industry than in any of the other regulated utilities.[2]

Price Controls

It is Section 40 of the 1986 Act that provides, crucially, for the designation of an airport. Designation (which is done by Statutory Order) imposes on the airport an RPI-X price-cap regime. Four airports have been designated, BAA's three London airports, as well as Manchester plc. These four airports are normally subject to a five yearly review of specified charges with the review of Manchester taking place one year later than the London review. The last round of reviews was conducted between 1995 and 1997. It is only selected 'airport charges' that are subject to explicit regulation (limiting the maximum amount that may be levied) and these are defined as those charges connected with the landing, take-off, and parking of aircraft, and with the handling of passengers through terminals. Individual charges as such are not subject to a price cap, but the latter is applied to the overall revenue yield per passenger; the airport operator, therefore, has a degree of discretion with respect to the level of each individual charge and the relationship between them (although, in practice, this discretion is limited by the pressures that airlines can, and do, exert on the airport operator through their powerful trade associations and sometimes through governments making representations on their behalf).

By statute, the regulator is obliged to perform his economic regulation function in a manner which furthers the reasonable interests of users (as well as promotes efficient, economic and profitable operation of airports, and encourages investment). Partly for this reason, but also because of past treaty obligations and custom in the international air transport industry, the determination of X during the quinquennial review is much influenced by what is commonly referred to as the *single-till* approach. Importantly, the regulator's judgement regarding an appropriate level for the price cap takes into account not only the revenue generated by airport charges but, in addition, the revenues generated by activities

---

2   Because of externalities associated with airports the planning control system is also onerous and the government is heavily involved in the decisions regarding the scale and location of new capacity.

such as retailing within terminals and the provision of rental property and other services to tenants and licencees (airline companies, car-hire operators, etc.) the fees and charges for which are not subject to a price-cap formula.[3] The revenue from retailing and property, which can dominate the revenues from price-capped airport charges (see Table 5.1),[4] are combined with the latter into a single-till as a prerequisite for judging whether the forecast of total revenue net of operating expenditure, provides a rate of return consistent with the firm's cost of capital and whether the projected cash flow is sufficient to sustain the prospective investment programme. The expectation of an inadequate return, or an insufficient cash flow, will lead to a less stringent price cap and vice versa.

Much of this price-cap review process is common to those undertaken for other utilities. However, the single-till in the airports' case, is an important difference. As Michael Beesley pointed out last year, whilst the other utility regulators have sought to isolate and bear down upon the natural monopoly elements, because of the single-till such focusing has not occurred in the case of airports.

Framework Reviews

Since the 1986 Act a couple of reviews of the legislative framework have taken place. The first, in 1994, essentially considered the *process* applied to economic regulation and resulted in proposals for 'streamlining' the approach, the most significant being an intention to adopt the standard utility model with a single economic regulator. Thus far the review process applied to airports by the 1986 Act has been unusual; it has been the MMC [now the Competition Commission] that has conducted the initial review and made recommendations to the regulator. This differs from the usual approach where the industry regulator conducts the review and a reference is made to the Commission only in the event of a utility challenging the regulator's conclusion.

The second review, in 1998, formed part of the government's general review of utility regulation. The thrust of this review was to continue aligning the process applying to airports with that applying to the other utilities. The proposals include: placing a primary duty on the CAA to further the interests of airport users; enabling the CAA to intervene when standards of performance have not been met or have led to disagreement with airport users; and providing the CAA with concurrent powers with the Director General of Fair Trading, under the Fair Trading Act of 1973, to refer possible monopoly situations to the Competition Commission for investigation (currently the CAA does not enjoy concurrent powers in this area). These changes will bring the CAA more in line with the other utility regulators.

---

3   These fees and charges have, nevertheless, been considered by the MMC during the quinquennial review process with a view to judging whether their level was contrary to the public interest.

4   Table 5.1 shows an *ex post* situation; the balance of revenue sources after the regulation of airport charges.

These proposals aim to tidy the edges of the existing regulatory framework. As a consequence, they are essentially conservative and they either choose to ignore, or fail to get to grips with, deeper problems inherent within the current framework. It is to these that I now turn.

Issues Arising from the Regulatory Framework

The Nature of the Airport User

One element in the package of proposed measures aligning airport regulation with the standard utility model, is the placing of a primary duty on the regulator to *further* the interests of the airport user. But this proposal contains a subtle, but potentially significant, difference from the government's proposal for the other utilities, which is to require regulators to *protect* the interests of consumers. Furthering interests is not the same as protecting interests and, if the object is consistency, it is not clear why the duty to be placed on the CAA is not also one of protecting the user.

This touches upon important developments in some of the regulated industries which have a bearing on how we might define the user in the case of airports. In the initial post-privatisation periods, most regulators were faced with a priority of setting price caps for services provided to millions of small customers. Subsequent de-regulation in telecoms and energy has meant price-cap regulation shrinking to focus on intermediate products (interconnection services in telecoms, transmission and distribution services in electricity and in gas). In these intermediate 'access' markets the demand side is characterised by much smaller numbers of relatively large buyers, some of which are, in global terms, much larger than the access provider. These are not 'consumers' in the sense usually meant by politicians. (Airport regulation is ahead of the pack in this area, in that the large user issue has been prominent from the outset, although it should be noted that the activities of the airport companies also impact direct on large numbers of individual travellers.)

Given these circumstances, the utilities bill, by placing a duty on the utility regulators to protect the interests of *consumers,* is arguably somewhat backward looking in its conceptual framework, and might also be criticised for its lack of clear thinking on the duties of regulators in respect of the relevant intermediate markets. It is, for example, by no means obvious that the primary aim of regulation should be to 'protect' or, even less obviously, to 'further' the interests of users, not least since, as competitors in downstream markets, there should be no presumption that such users have common interests. For example, in many, but not all, aviation markets there are the complications of bilateral air service agreements which often incorporate various barriers to entry.

In respect of congested airports, there is a further consideration. At an airport like Heathrow there are considerable scarcity rents which lead to rent seeking behaviour on the part of the different economic agents involved in the supply of air services. Such behaviour is constrained in the case of BAA by economic regulation but why should a regulator be given a duty to allocate a higher weight to the rents of users than to the rents of suppliers, such as BAA, which is what an unqualified application of the principle furthering the interest of users would imply? There are, therefore, important distribution issues to address which, in this case, take on an added significance, partly because of the size of the scarcity rents and partly because non-UK/EU users of the airport utility are a sizeable part of the market.[5]

Airport Designation Criteria

An airport to be subject to a price cap has to be designated under Section 40 of the 1986 Act and, as pointed out above, four airports are designated, the same four that were designated at the outset of regulation. Why these four and why others have not been added remains unclear, which is perhaps surprising given that designation is the crucial trigger for economic regulation. But economic regulation does not come without (sometimes considerable) disadvantages. Therefore, one might have expected airports to be treated on the basis of consistent and perhaps self-evident criteria when deciding upon which ones to designate.

In my 1994 lecture in this series,[6] I questioned why, at the outset, the Scottish airports of BAA were not designated. As a group, they dominated the Scottish airport market and in scale terms were similar to Manchester which was designated. Subsequently, the government reviewed the case for designating them and, at the same time and for the first time, set out criteria which it considered relevant to designation generally. These criteria include the extent of competition from other airports/transport modes and *prima facie* evidence of excessive profitability *or* abuse of monopoly position. To include both abuse of a monopoly position and evidence of excessive profitability as alternative criteria, is curious and appears contrary to the thrust of competition law with its focus on abuse. At what point profits become excessive is arguable and the danger is that any profits in excess of normal will be judged excessive. There is, after all, nothing inherently wrong with above normal levels of profit; such levels could indicate, for example, superior organisational efficiency and cost control and signal opportunities for profitable entry.

In the event, the government concluded that there was no case for designating the two principal Scottish airports, Glasgow and Edinburgh, because there was no evidence of abuse of monopoly position or inefficiency. However, both airports had achieved high levels of profit and rates of return. This suggests that, in spite of having included profitability in the list of designation criteria, in practice the government was inclined (correctly) to disregard it as a singular reason for designation. When announcing its decision, the government added that it believed the threat of designation provided a strong incentive for BAA to control its

---

5   We can note here attempts by US negotiators to shift the scarcity rents of Heathrow to US airlines, foreign ownership and control of which remains foreclosed.

6   Starkie (1994a).

charges. Possibly reflecting this, BAA did cap its charges on a voluntary basis (initially with the formula of RPI-3) at both Glasgow and Edinburgh.

This outcome begs the obvious question of whether this approach in Scotland is not a more appropriate way of approaching the economic regulation of airports generally. Rather than link the implementation of economic regulation to evidence that market power exists, (which appears to be the basis upon which the four airports were originally designated) would it not be preferable to hold reserve powers which are put into effect only when there is evidence that market power is being exploited and no voluntary agreement can be reached?

Service Quality

Whatever the merits of the current price cap applied to the major UK airports, it does mean that the focus of the regulatory system is upon charges *per se*. But, should service standards be brought within the ambit of regulation as well because of the incentives that price controls give for degrading service quality? If demand with respect to service quality is inelastic (i.e. a unit decrease in quality leads to a proportionately smaller decrease in demand) then it is possible that the cost savings from degrading the quality of service will exceed revenues foregone; profits will be enhanced as a consequence. For example, the quality of airside activities such as baggage handling, trolley services, holding lounges, can probably be degraded (perhaps severely) without any consequence for short-term passenger demand (in the long term adverse reputation may have an impact). On the other hand, demand for some commercial activities at an airport is probably (highly) sensitive to quality of service. Queuing by passengers will have a marked effect on demand for duty free goods and food and drink purchases, for example. Does this suggest that quality could be degraded in some activities to increase net profits, but not in others? Not necessarily. The willingness of passengers to undertake discretionary spending could very well depend on his/her treatment airside, so that this interdependence of demand may reduce incentives to degrade quality in *any* element of the airport service.

Although not subject to formal regulation of its quality of service by statute, BAA has for some time produced a quality of service index and provided specific service guarantees to airline passengers and to tenants; it also has consultation procedures with airlines and shares the results of market research with airlines, concessionaires and others involved in service delivery at its airports. The Monopoly and Mergers Commission has also taken into account complaints made by airlines and consumers when reviewing the price cap for the regulator.

Emphasis is now switching to the establishment of formal agreements between the airport companies and the airlines known as 'service level agreements'. After a period of experimentation, these are being introduced, although not without controversy. The main area of contention is whether the airport company should be penalised when performance falls below agreed standards. BAA considers that the processes covered by 'service level agreements' are shared between the airport and the airlines and that penalties should apply potentially to both parties. The airlines, on the other hand, believe that penalties should be payable by the airport alone; they argue that it is the airport that is the monopoly supplier whilst airlines operate in a competitive market and poor performance by an individual airline leads to passengers transferring to other airlines.

Clearly, from a practical point of view there are a number of difficulties if service quality is to be formally regulated. Judging the appropriate quality of service is difficult and it would be uneconomic to over-provide quality and inefficient to under-provide. The quality of service to aim for is that which would exist in a competitive market for the services in question. Bearing in mind that different airlines have different (sometimes very different) requirements, it is probable that in a competitive market there would exist a variety of service qualities attached to which would be different prices. The user would then choose the quality/price package which most suited its requirements. Unfortunately, the emphasis on service level agreements tends to focus the debate on a uniform (and possibly too high) standard and, thus, on a lower level of welfare than is potentially achievable.

The Structure of Charges

As Michael Beesley remarked last year, little serious attention seems to be given nowadays to peak load pricing. This is perhaps regrettable given the pioneering role in this field of BAA's predecessor, the British Airports Authority. The Authority introduced a policy of pricing peak demand at Heathrow and Gatwick in the early 1970s and, although the implementation left much to be desired [see Chapter 4], it was the first attempt to introduce, on a large scale, efficient pricing signals for the use of airports. The rudiments of the policy still exist, albeit substantially modified, and now focus upon runways and the utilisation of space for parking aircraft. There has also been an overlay of environmental charges whereby noisier aircraft pay more, sometimes much more, than quieter aircraft. The policy has come under fierce attack in the past from US aviation interests that felt that the peak passenger charge for the use of terminals unfairly penalised the early morning trans-Atlantic arrivals traffic at Heathrow.

What is particularly regrettable is that the structure of charges and its implications for economic efficiency, have not received more attention in the regulatory reviews. In fact, on the evidence of the three MMC reports reviewing Manchester, the Commission seemed to alter its position from positively encouraging the adoption of peak-load pricing (1987) to mild encouragement (1992) and then to indifference (1997), with no mention being made in the latter review of the earlier recommendations on the subject, in spite of their apparent neglect. It could be argued that by 1997 the cause was in fact lost; proposals to construct a second runway and provide substantial additional capacity were, by that time, well advanced. The time to have pressed the issue would have been in 1992 when a well-considered peak-load pricing scheme would have tested the case for a second runway and informed a decision on its optimal timing.

That apart, we have a situation at London airports where demand greatly exceeds supply at the current level of charges. There is a commonly held view that to clear the current London market requires a substantial hike in charges. This, of course, would run contrary to the current paradigm that results in a continual reduction in the real level of charge in spite of the congestion, so that, as a consequence, a balance between supply and demand is achieved through quantity rationing. But, there is an argument that *some* increase in charges would probably improve the situation even if it were not large enough to actually clear the market. This arises from the importance of the *structure* of charges regardless of whether the average *level* of charge clears the market. The underlying point is that total demand can be broken down into a number of subdemands (by routes, scheduled/charter traffic, transfer/ originating traffic, etc.) each of which will have its own elasticity of demand. With excess demand, quantity rationing will tend to lead to inefficiencies in the allocation of capacity between these submarkets. An increase in charges will, however, induce different substitutions among the sub-markets (there will be a greater reduction of demand where price sensitivities are higher). This will affect the rationing of the (now smaller) demand in ways that might be expected to improve efficiency a little. For example, if low value users are priced out, there will be more capacity available for higher value users. Thus, although the overall allocation will tend to be inefficient, there will at least have been some movement in the desired direction (i.e. some reallocation from lower to higher value users).

It may also be possible to devise an approach to charges, which clears, or more nearly clears, the market without greatly affecting the average charge. Suppose, for example, the passenger facilities charge (for the use of terminals) was abolished and the emphasis was placed entirely upon a fixed charge for landing aircraft (which might apply throughout the day if demand is at a constantly high level). This would bear more heavily upon smaller jets with fewer passengers, so that demand is reduced significantly in this part of the market, thus leaving runway demand and supply more balanced, but with an increase in the average size of aircraft.

A similar outcome in efficiency terms can be achieved by the introduction of a secondary market in take-off and landing 'slots' (entitlements for use of the runway at particular times). If incumbent airlines are free to trade their entitlements, the consequence will be that slots will be transferred to those airlines that place a higher value on their use. However, the allocation of slots is governed by European legislation and the Commission has so far opposed the trading of slots. This has not stopped a 'grey' market developing at both Gatwick and Heathrow, but in my view it would be better to legitimise the process and make it transparent. Nevertheless, a secondary market is not without its disadvantages; in particular it involves a lump sum financial transfer from entrants to incumbents with consequences for the balance sheets and financial strength of often competing airlines. Reducing the discrepancy between market clearing prices and the prices actually charged for the use of runways remains the preferred course of action.

The Single-till

I have already outlined the unusual but central role of the single-till in the overall approach to the price regulation of airports. It is an approach that has been criticised especially for forcing down charges at congested airports below market clearing levels and sometimes below the (resource) costs of providing airside services. Given the total revenue requirement, the amount of revenue from airport charges (and thus the average level of those charges) is determined by the anticipated level of profits from retailing and property activities. As more volume is squeezed out of a congested facility, retail revenues are increased and, if forecast to continue, the price-cap review will (all other things being equal) increase X, thus reducing charges at a time when economic efficiency requires these to increase. Indeed, if the price-capped airside activities contribute a minor share of total revenues, the gearing effect means that modest changes in revenue requirements give rise to large changes in (price-capped) charges.

This is the more evident distortion resulting from the single-till approach, but there are others. Although the retail and property activities are formally excluded from the scope of the price cap, nevertheless, the approach, by taking into account these revenues when determining the price cap, implicitly extends the range of activities subject to regulation. It means, for example, that the retailing activities are possibly subject to inefficient investment incentives. A further complication is that the overall regulatory approach requires the regulator to assess the airport company's cost of capital and an incorrect assessment will further distort investment incentives. The incorporation of retailing and property activities into the assessment inevitably complicates this exercise and increases the potential scope for error. Their incorporation requires, for example, the regulator to take a view on the cost of capital for retailing activities but this is not an area in which regulators, generally, have much or indeed any experience.

In spite of the distortions that the single-till imparts to the overall process, it is, nevertheless, an approach which focuses upon the important complementary nature of the relationship between the airside and retailing activities; an increase in the demand for flights from a particular airport will increase the demand for related goods and services and for rented property at that airport.[7] But there is a twist to this relationship that is particularly important. The retailing and property activities enjoy locational rents due to the fact that superior locations have an enhanced value (in just the same way that retailing properties in, say, Bond Street or Oxford Street enjoy similar economic rents from their own unique location).[8] And because the retailing and property activities gain these locational rents, increases in traffic volumes at an airport will often produce significant increases in their profitability.[9] For a profit maximising airport company with market

---

7   Reductions in *real* air fares can be expected to lead to additional demand for the complementary good because of changes in the cross-price elasticities, i.e. besides there being more passengers, a lower real air fare will encourage each passenger to spend more in airport shops.

8   As Martin Kunz (1999) points out (14) these locational rents are part of the process of allocating limited space efficiently.

9   These related activities are not always undertaken directly by the airport companies but are frequently franchised. The terms of the franchise normally enable the airport to extract much of the rent. Note however that Section 41 of the 1986 Act potentially constrains the ability of the airport companies to increase these rents by limiting the number of franchisees.

power, selling in both markets, the effect of the demand complementarity linked to the locational rents, is to attenuate the normal, downward pressure on profits which would arise when increased air traffic volumes have to be bought at the expense of lower prices. For example, in [Figure 5.1] a profit-maximising airport business, if considering only airside activities, will produce O_1 output and price at P_1 (left-hand diagram). However, increasing airside output raises demand for the complementary good (airport retailing), thus increasing location rents for the latter (right-hand diagram). A business with an interest in both markets will expand airside output providing that B+C >A, that is, up to Output O_2.[10] This means that, as long as an airport combines both activities, the incentive will be to set charges lower than if runways were a stand-alone facility.

This outcome has some potentially efficient properties. For example, if an airport is operating with excess capacity, efficient pricing of runways and terminals implies a level of charge that will fall short of cost recovery (including a reasonable return on capital employed). Absent the retailing activities, these efficient prices would not be achieved. If, however, there are economic rents from retailing activities, the complementary relationship will encourage the airport to increase these rents by lowering airside charges. That is, the returns from retailing may support a more efficient level of charges (although it is also possible that this situation could actually give rise to charges for airside activities that are below marginal cost).

Once the capacity of an airport is reached, however, the pattern of incentives will change. There is now nothing to be gained in not pricing the use of runways and terminals at market clearing levels; increased turnover in and profits from retailing cannot in the short-term be secured by reducing charges for runways and terminals below the level which equates demand with the capacity available. In these, capacity constrained circumstances, the prices charged for the use of airside activities by an efficient airport business which combines these with retailing, will be the same as the prices charged by an efficient airport business wherein runways and terminals are a stand-alone facility. Of course, once sufficient capacity is added, the situation will revert; the pricing of runways and terminals will take into account the contributory revenues from retailing and property.

Reform Options

More Focused Regulation

In the light of these important economic characteristics there are various options for regulatory reform that suggest themselves. The more obvious option would be to remove the single-till constraint whilst retaining price-cap regulation. In other words, maximum airport charges would be determined by a requirement to allow an

---

10   [This figure is repeated in *Chapter 13* as Figure 13.2, where a detailed explanation will be found in the accompanying text. A modified version of the figure will also be found in Australian Productivity Commission (2002) Appendix C.]

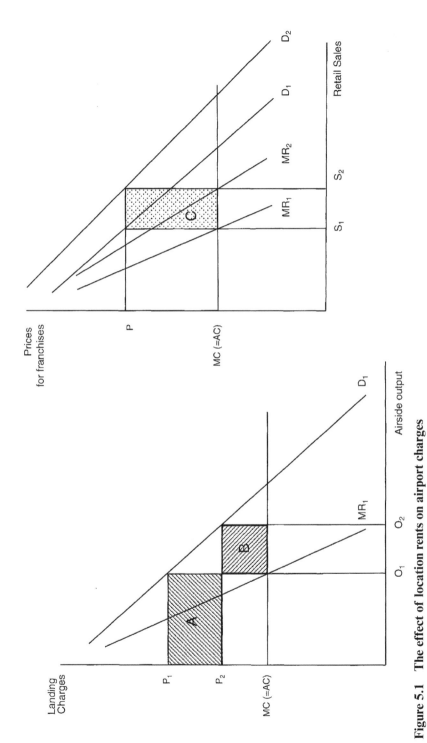

Figure 5.1   The effect of location rents on airport charges

Note: For an explanation, see text.

appropriate return on airside assets disregarding the revenues and costs of retailing and property activities. This might be referred to as more 'focused' regulation.

More focused regulation should bring about improvements in economic efficiency, since it would have the effect of reducing the distorted investment incentives previously outlined. In terms of its impact on charges, the effects here are contingent upon circumstances. At congested airports it is to be expected that average charges will increase, perhaps significantly, but, to the extent that charges have been held by regulation below marginal cost, the effects of the increase in economic terms will be beneficial. At uncongested airports it is more difficult to judge the impact of more focused regulation on charges. This will depend upon whether the existing price cap is actually binding on the airport company's charging decisions. The powerful economic incentives I have noted, whereby an airport company with significant earnings from retailing and with spare capacity has an incentive to keep charges lower than would otherwise be the case, means that the maximum allowable average charge may exceed what the company wishes to charge. If this is the case, charges will not increase once the single-till constraint is removed. Alternatively, the price cap may be binding, in which case charges will rise, but not necessarily to inefficient levels.

There will also be longer-term impacts. Removal of retailing and property from the regulatory assessments can be expected to affect both the pricing of, and investment in, these activities. As these sectors adjust, factors such as their price-cost margin will change, and these changes will have feed back effects on pricing and investment in these sectors. For example, if price-cost margins in retailing increase over time, the marginal profitability of extra demand will increase, and this will tend to increase incentives to price lower and invest more in (complementary) runways and terminal facilities. These incentives apply equally to congested and uncongested airports.

Overall, it would seem that from an economic viewpoint there is probably much to be gained by doing away with the single-till approach when undertaking the price-cap review. The risks from doing so are reduced substantially by the complementary nature of airport activities, and the incentives that these complementarities impart for efficient behaviour. However, if the regulatory framework was to be altered in this way it would be expedient to encourage airport companies to develop commercial activities that depend for their profitability upon the number of passengers using the airport. Less desirable would be moves to demerge, or hive off, these activities into companies separate from those owning and operating the runway and terminal infrastructure. Although a separate retailing or property company would continue to benefit from airport growth, the company retaining the airside assets may have less incentive to increase output. This is because the owners of such assets will no longer have a shared interest in the performance of the retailing and property assets (an interest which does exist if the company has an integrated structure). If the airport is operating at capacity this is perhaps of less significance but in the longer term it is likely to affect the incentives to invest. There might then be a requirement for greater regulatory scrutiny of the investment plans of the airport company to ensure that investment was at an appropriate level.

An alternative view is that, as a consequence of the demand complementarities, there would be incentives for separate companies to enter into contractual relationships that compensated the airport company for the beneficial effects that any expansion in its activities would have on the retailing company. To the extent that this was successful, we can note that all that would happen is that the effects of common ownership (integration) would be replicated by contractual arrangements, and the economic effects of demerger would be insubstantial. That is, demerger will only have significant effects if the contractual alternative is infeasible or costly (e.g. because of transactions costs) in which case these effects would not be beneficial.

Abolition of the Price Cap

A further option to consider is full de-regulation of both airport services and retailing activities. This involves the classic trade-off between the economic effects of market power and the distortions introduced by regulatory intervention. The option is, therefore, more/less attractive the greater/lesser the degree of competition among airports, and it raises associated questions concerning, for example, the common ownership of the three London airports.

But putting aside these questions for the time being, the economic factors that have led me to argue for the abolition of the single-till approach to regulation also lend weight to an argument for doing away with a permanent price cap. The existence of economic rents from complementary retailing and property activities has the effect of reducing some of the adverse effects of market power. It provides incentives for lower pricing and higher investment in airside activities because the extra air traffic will generate higher rents in retailing and property. The bundling of both types of activity is, therefore, a factor that, at least in terms of its effects on the market for runway and terminal services, tilts the balance of advantage towards de-regulation more than would be the case if runways were operated on a stand alone basis.

There is yet another factor associated with the economic characteristics of the industry which tilts the argument towards de-regulation. Compared with the more traditional 'natural' monopoly examples, supply in the airport industry is probably characterised by increasing, rather than decreasing, long-run costs at quite moderate levels of output. That is to say, if we double the output of a sizable airport by doubling the capacity available for use, total costs will more than double. This observation was made, somewhat in passing, in MMC2 (1991) and although there is not the hard, statistical evidence to support the proposition, it is, as I argued with David Thompson in 1985,[11] a likely outcome of the complex way in which airports grow in size. The source of the airport monopoly, therefore, is not the usual economies of scale in the long-run production function, but the fixity of 'locational' inputs (i.e. good sites) and economies of scope associated with established air service networks The significance of this increasing cost argument is twofold. First, even in the absence of congestion, prices in excess of average

---

11   Starkie and Thompson (1995).

costs are not necessarily inappropriate. And second, in increasing-cost industries, regulation of prices based on allowances for normal or reasonable rates of return on capital may lead to inefficiently low prices. Thus, even though de-regulation may, in spite of the moderating factors previously mentioned, lead to inefficiently high prices, the outcome is not necessarily worse than the regulatory outcome.

In addition to these economic arguments supporting the case for de-regulation, there is a further argument which derives from the earlier observations made regarding the nature of the airport user. As we observed, the airport user from a regulatory standpoint is not so much the individual airline passenger but rather a relatively small number of airlines providing services in downstream markets. These airlines, when represented through their trade associations, are relatively large and sophisticated customers more than capable of challenging the airport operators. They have often done so and frequently have used the courts to put their case. Arguably, therefore, the airlines have a measure of countervailing power. However, the countervailing power of airlines would be greater if the London airports were all in separate ownership. For example, it is likely that Ryanair's recent threat to leave Stansted unless a proposed increase in charges was ameliorated, would have carried more weight if Stansted had not been in common ownership with Gatwick and Heathrow.

This suggests that there is a trade off worth contemplating between abolishing or reducing the level of price control and separating the ownership of proximate airports. To achieve the level of restructuring required to advance airport competition and to ease the regulatory burden might not be too difficult. As the voluntary divestment of generating capacity in the electricity supply industry and the voluntary separation of Centrica from British Gas indicate, it does not necessarily require considerable political will (in the gas case the government had rejected the MMC's divestment proposals but the company decided that restructuring was in the best interests of the shareholders). If BAA were confronted with the possibility of a much-reduced regulatory burden in the event of divestment, it might be a trade-off that the company was willing to take seriously.

Taking into account the high level of existing demand relative to available capacity at both Heathrow and Gatwick, in the event of separate ownership, competition will centre less on prices and more on the second of the two competitive dimensions set out by George Yarrow in Chapter 3. Separate ownership in this airports' case is more likely to spur product development and innovation especially with a view to adding more capacity to the congested infrastructure. Although, as I have argued, BAA currently has an incentive to pursue the same end, competition is likely to sharpen creativity. Consequently, it might be argued that for the capital markets there might be little to loose by BAA being broken-up into potentially competing parts (the sum of the profitable parts may even exceed the sum of the existing profitable whole) and everything to gain from a less regulated environment.

The principal difficulty with de-regulation is likely to be associated with distributional issues (rather than issues of economic efficiency). At *congested* airports, removal of the price cap can be expected to lead to user detriments in the form of higher prices to users and to significant increases in the profits of the airport companies. Consequently, even if efficiency is improved (if profit gains to the airport company exceeded loss of profits to incumbent airlines and loss of consumers' surplus to passengers), the weighting given to the user interest may be decisive and make de-regulation difficult, if not impossible, to achieve. Increased competition between airports could help to mitigate such distributional effects, but, if de-regulation was to be pursued as a serious option, it might be necessary to consider it alongside other measures designed to offset the adverse effects on users.

One such measure would be to allow the airlines take a stake in the airport. In Australia and New Zealand, for example, some domestic airlines own their own terminals and have thus vertically integrated some of the service functions typically provided by the airport company. However such developments provide opportunities for foreclosing entry into the air services market and new entrants, particularly in Australia, have faced this situation. Although access rules could be formulated to try to ease this problem, an alternative (structural) approach would be to demerge airport *retailing* into a separate company which is then established as a joint venture between the owners of the runway and terminal assets, and the (downstream) airline companies.[12]

A joint venture of this type would maintain existing incentives for the airport company to increase output and, in addition, provide incumbent airlines with similar incentives which they do not necessarily have at the present time (given the strategic competition among airlines for restricted landing slots at congested airports). Such a company structure could provide a means for compensating the airlines for losses following de-regulation. Whether the required amount of compensation could be achieved would depend upon the willingness of the airport company to provide incumbent airlines with enough equity in a joint venture. This might be possible. Increased charges at congested airports could be expected to lead to an increase in the average number of passengers per aircraft movement and, because of this, there would be an increase in total retailing turnover and profits which both parties to the joint venture could share; there could conceivably be a net overall gain to the airport company after allowing for the lump sum transfer of enough equity to compensate the airlines for any increase in charges.

Conclusions

Overall, I believe the balance of the argument is against a continuation of the single-till approach to the regulation of airport charges. In the longer term, the approach introduces distorted investment incentives particularly by extending the

---

12   The 'deal' put forward last year by Michael Beesley was essentially one between the government and BAA (with the airlines offered the comfort of charges being pegged in real terms). However, in para. 49 of his paper he also hinted at the possibility of a joint venture between BAA and the airlines in relation to the development of Terminal 5. It would seem, though, that he viewed this 'joint venture' in terms of airlines bidding for access rights to the new Terminal.

scope of regulation; it requires the regulator to make difficult judgments regarding the cost of capital not only for the air transport sector but also for retailing and property. In the shorter term, at congested airports, it also leads to distorted pricing signals. These distorted economic incentives are probably reason enough to abolish the single-till approach. But a more telling reason for its abolition is that it is an unnecessary complication. A business that combines the landing of aircraft with retailing will have, without the intervention of a regulator, an incentive to reduce charges to airlines and to expand output. The special factor in this favourable situation is the union of strong demand complementarities (the demand for the use of the runway and the demand for retailing and property facilities) with location rents (from retailing and property).

The added significance of the demand complementarities combined with economic rents is that they also make it less likely, and possibly *much* less likely, that any market power that does exist in the market for landing aircraft and handling passengers will be abused. In fact, because the airport company has incentives to maximise throughput, the problem is more likely to be one of excessive use and congestion (rather than the classical problem of a monopolist restricting output).[13] It is important, therefore, not to jump from the proposition that the operator of, say, Heathrow has significant market power (which is certainly the case) to the proposition that the operator can be expected to abuse that market power. Consequently, the special economic circumstances in the airport case tip the balance of the argument not only towards the abolition of the single-till approach to price-cap regulation, but also towards the abolition of price-cap regulation *per se*.

In addition to the underlying pattern of economic incentives there are other aspects to take into account, that lead to the conclusion that serious consideration should be given to removing the formal price cap at airports currently designated. Foremost amongst these is the new Competition Act, which will give the CAA considerable powers to police and punish abuse of a dominant position thereby establishing ex-ante incentives not to abuse market power. With this in mind, if designation was set aside, it might be appropriate if this was accompanied by undertakings in respect of pricing at Heathrow (and possibly Gatwick and Manchester) given under general competition law; such undertakings might have an economic effect similar to the imposition of a relatively loose price cap at Heathrow.

The overall thrust of such moves would be to shift airport regulation in the direction of the approach adopted in New Zealand, where the privatised airports are subject to reserve powers of price control under the Commerce Act, and, importantly, towards the approach now adopted for BAA's main Scottish airports. At the latter, charges are capped on an informal basis following an agreement with the CAA. This establishes an important precedent, broadly along the lines of the approach I am suggesting here.

Finally, there is the question of timing if radical change is to take place. One necessary step is to notify the US authorities of any proposal to abolish the single

---

13   Forsyth (1997).

till, as is required under the 1994 amendment to Article 10 of the Bermuda II Agreement. (Article 10 was amended so that there is no longer a requirement for the UK to maintain a single-till approach). A further consideration is how change can be introduced within the timetable of the regulatory review process. The next review (of BAA) was due to take place, starting next year, with completion expected in 2001, but the timetable has been put back a year to allow time for a decision on [Terminal 5 at Heathrow] and to allow for completion of the general review of aviation policy that the present government has set in train. This delay provides an opportunity to undertake an appraisal of the regulatory framework as a whole, the outcome of which may make a further quinquennial review unnecessary.

Chapter 6

Regulatory Developments[1]

Introduction

In the 2002/3 [CRI] *Regulatory Review*, Ralph Turvey had the very challenging task of reviewing the final year of a long process leading up to the setting of the current price caps for both Manchester and BAA's London airports. The process had started in July 2000, when the Civil Aviation Authority (CAA) decided to undertake a fundamental review of the regulatory approach. This was followed in early 2002 by a referral, required under the 1987 Airports Act, to the Competition Commission. The Commission then made recommendations to the CAA prior to the latter's final proposals and subsequent decisions in February and March 2003; these took effect from April 2003.

Ralph Turvey was able to set the scene right up to the CAA's final proposals, but he was not quite able to close the chapter with the final decisions on Manchester and BAA, both of which were made after the last Regulatory Review had gone to press. In the event, the CAA's two decisions differed from its proposals only in some minor ways. And, as Turvey pointed out, the CAA's Proposals generally accepted the Competition Commission's recommendations; the only difference of real significance being the future treatment of each London airport on a stand-alone basis rather than treating them as a single airport system. This means that it will no longer be possible for BAA to leverage its market power at Heathrow to support further development at Stansted.[2]

The final airport price caps allow Heathrow to *increase* its charges in real terms by 6.5 [per cent], per annum, thus reflecting the considerable capital spend on Terminal 5 during the 2003/8 period. Two new elements were introduced into the Heathrow cap: first, a 'trigger', which reduces the maximum allowable charge when the airport has not achieved particular capital project milestones on time and, second, an incentive for BAA to increase peak period runway capacity. At Gatwick (also subject to a trigger) and at Stansted, the price cap was set at zero. (This cap was expected to be marginally binding at Gatwick but non-binding at Stansted in view of the latter's lack of pricing power). At Manchester charges are to fall in real terms by 5 [per cent], per annum.

---

1   First published as 'Airports Regulation' in P. Vass (ed.) *Regulatory Review 2004/2005*, CRI, University of Bath and republished here by kind permission of CRI.
2   For the significance of this change in policy, see Starkie (2004).

The Regulatory Framework

The path to the decisions had been a long and, some would say, winding one. A very large quantity of the material had been produced by all sides and yet the basic regulatory approach (save for the setting aside of the system approach in the case of the London airports) had changed little; radical features proposed initially by the CAA (such as the dual-till approach and proposals for a long-term price path) did not find favour with most airlines nor a rather conservative Competition Commission and the two bodies were often at loggerheads; the final convergence of view was rather more apparent than real.

Inevitably this led, subsequently, to criticism of the overall process. BAA has been particularly acerbic, commenting that:

> ... they [CAA and the Competition Commission] took contrasting policy positions on virtually all the issues of principle, and the process led to no meeting of minds ... on key issues .... A more regrettable aspect of the dual review process, which goes to the question on excessive regulation, was the failure of the two regulators to manage and co-ordinate on the bread and butter analytical work.[3]

And:

> Where two regulators can differ as much as the CAA and Competition Commission appear to, the credibility of regulation itself is open to question.[4]

Of course, as Ralph Turvey reminded readers last year, the regulatory structure for airports is unusual. It differs from the other regulated industries because the Competition Commission has been given a mandatory role as adviser to the CAA. In other industries, the Commission becomes involved only when a regulated utility refuses to accept the regulator's decision. In 1998, the government sought views through a formal consultation as to whether the CAA should become the single regulator for airports, with a mechanism for referral to the Competition Commission only in the event that there was a disagreement between the regulator and regulated companies.[5] Following the consultation, the government confirmed its intention to implement this and other changes but it did not include them subsequently in the Utilities Act. In the light of developments outlined, this was perhaps unfortunate.

CAA Consultation on Regulatory Processes

Since the conclusion of the 2003/8 price-cap reviews, the singular event of some importance has been the CAA's launch of a consultation that looks at lessons to be

---

3   Toms (2003).
4   Toms (2004).
5   Department of Trade and Industry (1998).

learnt from last time round and sets out in some detail how it believes that airport regulation should develop in the run up to the next reviews.[6] Notwithstanding the fundamental review at the time of the last price-cap reviews, the consultation document includes itself some radical proposals, but their focus is essentially on the 'processes' involved in a price-cap review, rather than on methodological issues.

The document acknowledges that the last airport reviews were ambitious in scope and covered a significant number of weighty issues. This, in turn, created a degree of uncertainty during the review process, and meant that the process itself was resource intensive for all concerned. It also acknowledged that a number of policy issues were highly contentious. For the future, the CAA has stated its intention of adopting a more evolutionary approach so that fundamental regulatory choices are not revisited every five years. But, its key proposal is that in future the CAA should act more as a facilitator to enable the regulated companies to engage with the airlines and other users in a more structured and hopefully constructive dialogue. It is felt that airport companies and airlines are better placed than the CAA to understand user needs and are, therefore, in a better position to reduce regulatory risk by taking more control of the regulatory agenda through embedding it in their own commercial processes.

Specifically, the CAA is suggesting that the business plans for each designated airport should be drawn up following detailed consultation with the airlines and that this process should precede, for the purposes of the price-cap review, submission of the plans to the CAA. It is intended that the airports and airlines will agree the key variables of capital expenditure (CAPEX), operating expenditure (OPEX) and service quality that feed into the submission. The CAA will continue to take the lead on issues such as the cost of capital, where it considers it has a comparative advantage. In the absence of agreement the CAA will, as now, provide a regulatory determination: but recourse to the regulatory decision-making process is to be seen as a last resort.

In principle, this proposal by the CAA is appealing; it holds out the prospect for a lighter regulatory touch in general, for less regulatory risk, and a lower cost of capital. But there are some major hurdles to overcome if the new approach is to be successful.

First, it is arguable whether too much is expected from airlines in particular. At one time they were considered the apogee of a contestable industry with very little sunk costs ('capital on wings' was the coined phrase). Although a less sanguine view is now taken of whether the airline industry is fully contestable, there remains a sharp contrast between it and the airport industry with the focus of the latter on capital with generally a long economic life, irretrievably committed to a specific purpose in a specific location. Consequently, the planning horizons of the two industries differ substantially: added to which is an imbalance of resources and expertise; some airlines have no staff versed in airport or general utility regulation, whilst others have only limited resources devoted to such issues. It is a problem that potentially could be eased by the airlines in time developing their in-house

---

6   CAA (2004a).

expertise or, alternatively, by using airline trade associations to pool resources. But, importantly, different airlines have different agendas; the expectations and requirements of low-cost carriers such as Ryanair or easyJet are rather different from those of the full service carriers like BA, and the requirements of the two groups are sometimes mutually exclusive.

A second difficulty concerns the role of the Competition Commission in the overall process and the CAA's proposal that, in future, the Commission should not undertake a full regulatory review but focus only on those areas of continuing dispute. Until now, the Competition Commission has not shown a willingness to disengage; it has tended to cover more and more issues in increasing detail at each regulatory review. It has done this most probably to avoid the risk of a legal challenge to its recommendations and it is difficult to foresee this situation changing, especially so if the airlines seek to influence the Commission in the event of disagreement with the regulated airports on various issues. In anticipation of such differences the CAA is signalling that it will reach a view on the merits of the case and not split the difference between entrenched positions so that the parties need to consider – in addition to the perceived upsides – the potential downsides of leaving decisions to the regulator. But there is no requirement for the Competition Commission to follow suit and it remains to be seen how in the future the CAA will handle disagreements between the two regulators; although the CAA has the final say, last time it appeared to be the one to concede most ground perhaps mindful itself of a legal challenge.

Following consultations on its May 2004 document, in December the CAA announced that it proposed to publish a further consultation document in March 2005, setting out in more detail a framework to enable discussions to take place between airports and airlines.[7] In doing so, it signalled its intention to press ahead with its earlier proposal but, it noted: '[T]here are undoubtedly some significant issues to be resolved on which there is not – nor would we expect there to be – universal agreement.' Amongst these issues, the degree of information to be provided by the airports to the airlines in a timely manner and the need for both parties to engage at a senior level were highlighted (the latter echoes the disparities in expertise noted above). Finally, the CAA commented that, having observed recently the interplay between airports and airlines at a number of the designated airports, '… the nature and depth of … disagreements appear to be very substantial'. It added that, unless this was a passing phase, it would call into question whether the proposed approach based on increased airport/airline engagement should be applied at all the designated airports or whether at some a more traditional regulator-led review would be more appropriate.

De-regulation?

The CAA's proposed approach is in the spirit of trying to reduce the extent of regulation, which during the last price-cap review was widely acknowledged as a

---

7   CAA (2004b).

burden to all concerned. But, regulatory creep is a powerful force and, as Toms has pointed out:

> The first review [of BAA], in 1991 took 12 months. The second review in 1996 took 21 months, the ... 2002 review [took] 32 months. The first formula was a simple value of 'x'. The current proposed formula is a value of x, with six trigger points, and a service quality scheme covering a dozen parameters. In addition, the BAA operates under a web of agreements and undertakings, which have grown at each review. (Toms, 2004)

Perhaps the time has come, therefore, to place on the agenda once more the question: is it still appropriate to subject all four airports designated in 1987 to a continuing regulatory review.

Manchester has always been in an anomalous position in so far as it is wholly within the public sector and, as Turvey remarked in the 2002/3 Regulatory Review, the need for economic regulation of a local authority airport is far from obvious. Nor is it obvious that, if economic regulation is needed, standard price-cap regulation is the appropriate form to apply.[8] Price-cap regulation is designed to incentivise a profit-maximising firm, but Manchester admits to pursuing a wider agenda and is not seeking to maximise profits *per se*.

The broad purpose of economic regulation is to remedy an insufficiently competitive marketplace but Manchester, today, faces more effective competition from other airports than it did when it was first designated nearly twenty years ago. In 2005, the competitive environment Manchester faces will increase with the opening of commercial flights from a former RAF base near Doncaster. Consent for this development was given in 2003 (in spite of objections from Manchester). It poses a major threat to Manchester because of its long runway, located within a region from which Manchester has previously drawn much traffic.

Stansted is another anomaly as a designated airport. It was first designated when it had a tiny annual throughput of 0.5m passengers. In spite of its rapid growth in recent years (it is now larger than Manchester), it is viewed as having relatively little market power and it has struggled over the years to price up to the notional price cap set for it. In the past, de-designation would have been difficult because of the integration of Stansted's regulatory asset base with that of Heathrow and Gatwick as part of the system approach to the regulation of BAA's London airports. However, with the change of CAA policy following the last review, with each airport now assessed on a stand-alone basis, this constraint on de-designation no longer applies.

De-designation of Manchester and Stansted would leave just Heathrow and Gatwick to be subject to the existing regulatory framework. This narrower focus would reduce the regulatory burden on the airport industry and on airlines, particularly those using more than one designated airport. The opportunity might also be taken to de-couple the price-cap review of Heathrow and Gatwick, so that

---

8   Starkie (2003a) suggests that more stress is placed on the structure of prices rather than the level of average prices.

they are sequenced a year apart.[9] Moreover, all this can be accomplished without the need for primary legislation; the 1986 Airports Act provides for designation, and thus de-designation, by the Secretary of State using secondary legislation and Section 20 of the Statute also allows for the deferment of a review by up to one year.

---

9   As part of its new proposals, the CAA is also considering sequencing the review of Manchester *vis* BAA, thus returning to the *status quo ante* that preceded the 2002/3 Review.

PART III
Economic Regulation:
Some Issues

Overview

The method by which airport charges have been regulated in the UK – the application of an RPI-X formula – has in recent years been widely adopted for the regulation of utility industry charges elsewhere in the world. This includes the United States where previously rate-based (rate-of-return) regulation held sway, but applied originally to the three London airports of BAA, there was an unusual feature; the price caps for the three airports were interdependent.

This interdependency was achieved by the particular treatment of BAA's assets in the regulatory process. The value of X was calculated to allow for a reasonable return on assets (after allowing for other factors such as anticipated traffics, operating costs, etc.) and this rate of return was calculated on the *combined* assets of the three airports; if any one airport was unable to earn its cost of capital, prices were adjusted at the other airports so that the group as a whole earned its cost of capital. This approach, which has the effect of allowing sub-commercial charges at one airport to be supported by higher charges elsewhere, is known as the *system approach*. It is, of course, an approach made possible by the joint ownership of airports; by subjecting all three airports to a price cap; and by a permissive regulatory approach.

The next three chapters are concerned, directly or indirectly, with the system approach and specifically with the way that the controversial large-scale development of Stansted in the late 1980s was made possible by the approach and by its implicit cross-subsidisation of the expansion project. At the time of the debate on the 1986 Act, concern was expressed by non-BAA airports that Stansted might be cross-subsidised in this way. But the government tried to allay those fears. *Chapter 7* is a short piece written with David Thompson, published early in 1986. It takes a very sceptical view of the government's emollience, in part because we had 'run the numbers' on the expansion project and had found its commercial viability to be suspect; *Chapter 8*, first published in *Fiscal Studies* (Starkie and Thompson 1986a), explains this finding. It presents calculations which indicated that the immediate development of Stansted on the scale proposed was unlikely to show an acceptable rate of return: the public sector's own calculations had remained confidential. History has since confirmed our judgement; even today, more than 25 years since the expansion planned in the 1980s was completed, Stansted still struggles to price at a level that provides an adequate return on assets.

Recent years have seen, however, a fundamental change in the regulatory approach. As Chapter 6 pointed out, the relationship between the CAA and the Competition Commission (CC) during the review setting the price cap for 2003–08 was rather difficult. Most of the more radical proposals put forward initially by the CAA were not supported by the CC and the CAA did not press its case when it came to the final Determination. There was, however, one, very important, exception and that was the CAA's proposals to do away with the system approach and to set the price cap for each London airport by taking into account its separate (regulatory) asset base. This, so-called, stand-alone approach came into effect in April 2003. The implications of this change were huge because at the time the BAA was preparing for a further substantial expansion at Stansted, for which it was considered necessary to have the support of a system approach.

Chapter 9 provides an overview of the political economy associated with the expansion of Stansted in the late 1980s. The chapter starts with the 1985 White Paper on airports policy and ends with the December 2003 White Paper that set out the strategic framework for the future development of airport capacity in the UK. The chapter has the object of placing developments in a framework that examines certain aspects of price regulation, hence the title 'Testing the Regulatory Model'. In this context a point made is that the system approach provides the regulated company with an opportunity to adopt an entry deterrence strategy by leveraging its core market power through indulging in excessive investment in a sub-market subject to possible entry.

In 2004, shortly before the original paper was published in *Fiscal Studies* (Starkie 2004), the restriction in the BAA's Articles of Association limiting a single shareholder to no more than 15 per cent of the equity was removed; subsequently this allowed a Ferrovial-led consortium to acquire BAA plc in June 2006. Early in 2007 BAA announced significant changes to its expansion proposals for Stansted, resulting in the total cost of the project falling from £4.00bn, envisaged at the time of the 2003 White Paper, to £2.27bn (albeit for a slightly smaller increment of capacity and a later completion date) and in the cost of the first stage falling from £1.73bn (in 2005 prices) to £1.42bn. It is probable, therefore, that greater capital market disciplines, together with the prospect of a continuation of the stand-alone approach, had conspired to produce a more capital efficient outcome whereby capacity was added at a lower unit cost.

At the time of its 2002 review, the Competition Commission anticipated that Stansted would increase its net yield gradually to match that of Gatwick. There has been no evidence of this occurring. In 2005/6 the average yield per passenger was only two thirds of the corresponding Gatwick figure (Table A-6/7, 247, CAA 2006); Stansted continued to exercise little market power. The issue of de-designating Stansted has, however, reached the policy agenda with the launch in 2007 of a consultation on the issue by the Department of Transport. Should, as a result, Stansted be de-designated, it would of course no longer be subject to a price cap. Because the price cap (set equal to RPI from 2003) has not been a binding constraint, such a decision would simply accept the realities of the market. But, the real significance of de-designation is that it would prevent the distortions and unintended consequences inevitably associated with economic regulation, a point explicitly recognised in December 2006 by the CAA (see para. 23.14, CAA 2006).

The next two chapters address a closely related theme, namely, investment incentives. When the utility industries in the UK passed into private ownership, the initial limit on prices was set by the government and these limits were to be subject to periodic review. In the case of the BAA airports, the first review started in 1990. It was readily apparent at an early stage that investment was a difficult issue. At the time, the very large Terminal 5 project at Heathrow was in the pipeline and the importance of the price-cap decision for the return on this investment was recognised. But the project's return also depended upon permitted charges over a number of review periods and at each review the Regulator could take a different view on fundamentals such as the firm's cost of capital; the 'hold-up' problem had become, therefore, an issue. *Chapter 10*, first published in the *Journal of Transport, Economics and Policy* (Starkie 1991) describes the ensuing debate on the implications of the 1992 price cap for investment in new capacity.

The 'hold-up' problem is one factor that academic commentators have referred to when arguing that the price-cap approach can discourage investment in new capacity and encourage the 'sweating' of assets. *Chapter 11* reviews these arguments and then considers three counter-arguments that suggest that the regulated firm does not have an incentive to under-invest. The first of these has since become very apparent at Heathrow airport where in the summer of 2007, a combination of security concerns and a capacity squeeze prior to the opening of Terminal 5, has lead at times to chaotic conditions; it is the argument that under-investment risks the reputation of the company and risks imposing significant additional costs on the management of the firm,[1] as well as leading to the possibility of more intrusive regulation. A second argument reflects the agency issue; it is that management might seek to fulfil its ambition through the scale of its investments (see also Tirole 2006). And third, excessive investment might form part of an entry deterring strategy (the point made in Chapter 9 in relation to Stansted airport). The chapter then briefly considers the investment record of the price-capped airports, including those in Ireland. On balance, it concludes that there is little evidence that the hold-up problem has limited the scale of airport investment and that allowing the regulated company its cost of capital, on an investment programme that regulators have generally chosen not to second-guess, has probably lead to inefficient investment.

The final chapter, *Chapter 12*, in this part of the book is concerned with the issue of the revenue till: should the price cap be fixed with or without reference to the revenues that an airport generates from its commercial activities – retailing, renting office space, providing car parking, etc. The issue was first raised in Chapter 5, and the argument therein basically summarised the main conclusions of a commissioned report on the issue for the CAA that I had prepared with Professor George Yarrow. This report can be found on the CAA's website. In the event, the CAA in its review leading up to the 2003 Price Determination agreed with our conclusions and favoured the abolition of a mandatory single-till. This proposal did not find favour, however, with the Competition Commission and the Commission went to considerable lengths to argue in favour of the *status-quo*; not only did it devote a substantial part of its final report to the theme, but

---

1   I have in mind here additional stress and general 'hassle' faced by management, quite contrary to notions of a 'quiet life', hence my use in the chapter of the phrase 'internal costs'. The firm, in addition, will be faced with increasing short-run average costs as existing capacity is increasingly utilised.

it unexpectedly issued an interim report on the single-till and invited comments. Chapter 12 is my response, an edited version of which was included in the Commission's Final Report. The Commission was not persuaded; consequently the CAA decided to maintain the mandatory single-till approach in the 2003 Determination and it has adopted a similar position thus far in relation to the 2008–13 price-cap review.

Chapter 7

Airport Cross-Subsidy[1]

During last year's debate [1985] on the White Paper on airport policy, MPs from both sides of the House expressed concern regarding the degree to which Stansted was cross-subsidised by London's other airports. The Secretary of State, Nicholas Ridley, went to great lengths to allay this concern: cross-subsidy would be 'cut out' by:

- establishing each of the British Airports Authority's airports as a Companies Act company (wholly owned by a holding company, BAA plc); and
- ensuring that loans from the holding company, or between separate airports companies, were at commercial rates and transparent.

However, those White Paper proposals will not become effective until after privatisation (probably not until the spring of 1987). Meanwhile, Ridley has to reach a decision on the Authority's investment submission for Stansted, which arrived on his desk before Christmas. Its approval is the final hurdle that the Authority's chairman, Sir Norman Payne, faces in achieving his aim of creating the 'best airport in the world'.

It is quite clear from the White Paper that the government expect the Authority to show that its investment in Stansted would earn an acceptable rate of return. But there must be serious doubts whether the current proposal, for an immediate investment of between five and six million annual units of terminal capacity, is capable of doing that. Our calculations, based on the figure of £270mn mentioned by Ridley in June last year, indicate that Stansted would provide a satisfactory return, at current real average revenues per passenger, only if the investment is written off over 25 years – which is long for the private sector – and current operating costs are considerably reduced very significantly [...].

If Ridley does approve an investment on those terms, it will mean that the Authority will be placing contracts long before privatisation. The new Board, whatever its views might be, will find itself locked into a commitment to expand Stansted in a particular way. Discretion on the pace and the type of investment will no longer be available.

Of course, that need not concern a privatised British Airports Authority. Insofar as the capital markets perceive a big one-off tranche of capacity at Stansted as a poor investment, the effect will be to reduce the sale proceeds of the Authority

1   Co-authored with David Thompson and first published in *Public Money*, March 1986 (now *Public Money and Management*) and republished here by kind permission of CIPFA.

on privatisation. In other words, the extended risks associated with too rapid an expansion of capacity will be borne largely by the taxpayer, not the company.

However, it should concern the government if they genuinely want to avoid cross-subsidy. Over-provision will place the Civil Aviation Authority, which is to have a new role policing predatory pricing by airport authorities, in a difficult position. With excess capacity in place, low charges and cross-subsidy from Heathrow and Gatwick are legitimised. Plus ça change ...

The government might, of course, argue that by cutting the private sector out of the Stansted decision, they avoid the risk of the private company not investing at all, and simply sitting back and earning easy profits at Heathrow and Gatwick. But if they were afraid that a private company would act against the public interest in this way, should they be privatising the British Airports Authority in the first place?

Chapter 8

Pre-empting Market Decisions[1]

Introduction

The proposal to develop a third major London airport at Stansted has long been controversial, both to residents around the site and to advocates of the case for the development of facilities at regional airports in preference to London. In its White Paper last summer the government gave initial approval to the development of a major new terminal at Stansted. It stated, however, that it 'will expect the BAA to show that the investment will earn an acceptable rate of return' (Secretary of State for Transport, 1985, para. 5.27). Also in the White Paper – and the ensuing legislation now passing through Parliament – the government announced its intention to privatise Stansted's owners, the British Airports Authority (BAA).

In the event, the decision on whether to go ahead with Stansted has *not* been deferred until BAA has passed into the private sector. Final approval to go ahead was granted by the Secretary of State on 10 March 1986. Provided that the investment meets the criterion set by the Secretary of State – of earning an acceptable rate of return – then proceeding immediately with the investment should strengthen BAA's standing as it is prepared for flotation. There are grounds for believing, however, that BAA has in the past tended to press for the premature expansion of its capacity (see Starkie and Thompson 1985). If this is true in this instance then the immediate development of Stansted will both weaken BAA's potential stock-market valuation and, importantly, provide it with an additional advantage in attracting traffic in competition with other airports.

The critical issue is thus whether the immediate development of Stansted will show 'an acceptable rate of return'. Unfortunately, the long tradition in the UK of public sector investment appraisal (following Treasury guidelines) does not extend, except in a few instances, to publication of the results. It is, however, possible to reconstruct the main elements in an evaluation of the Stansted investment. Such an exercise can clearly only be indicative; full information is not available. However, the results of the reconstruction appear to be reasonably robust. And they throw interesting light on one facet of the government's privatisation programme.

---

1   First published in *Fiscal Studies*, 1986 and republished here by kind permission of the Institute for Fiscal Studies.

The Stansted Project

BAA proposes to construct initially a terminal capable of handling up to 8 million passengers per annum (mppa), together with the associated infrastructure necessary to support this volume of traffic. At present, Stansted has a terminal capacity of about 2 mppa which will be replaced by the proposal; the net addition to present capacity will thus be 6 mppa.

The Secretary of State (1986), when approving the investment proposal in March, confirmed that the traffic forecasts used by the Authority in its financial appraisal were consistent with those set out in the White Paper on airports policy published in June 1985. It is proposed to open the new terminal late in 1990 by which time existing capacity is expected to be fully used. Capacity of the new terminal is expected to be reached in 1995, four years after it is commissioned.

The estimated cost of the proposed development is £290 mn (at October 1985 prices). Expenditure is to be spread over nine years with £33 mn committed for 1986/87. The Secretary of State added that 'this investment should meet the required rate of return for public sector investments'. But during the Committee stages of the Airports Bill he declined calls to publish further details of the appraisal.

In spite of there being only limited information on a number of points, enough basic detail is now known to enable a review of the proposed investment's viability. To undertake this review we have made a number of assumptions. These concern:

- the precise phasing of construction expenditures;
- the running costs of the existing and proposed terminal; and
- the net income to be expected from the new facility.

Throughout we have discounted all values to 1985 using a 7 per cent rate of discount. In order to ensure the achievement of an overall rate of return of 5 per cent as required by a White Paper on Nationalised Industries (Chancellor of the Exchequer, 1978), the government seeks an initial estimated return of 7 per cent as a margin for risk and appraisal optimism (see MMC 1985). We have considered two alternative profiles for the timing of expenditure on the development (see Figure 8.1). Because the terminal is expected to be open by 1991, the bulk of the associated aircraft aprons, taxiways and access roads will also be required by this date. Consequently, the scope for slipping expenditure beyond 1990 is limited. present value terms) of £239.3 mn.

In profile A, expenditure is loaded toward the early part of the construction period and this represents expenditure (in 1985 present value terms) of £239.3 mn. In profile B expenditure is loaded toward the later years and this represents an expenditure of £236.6 mn (also in 1985 present value terms). Given the constraint of the opening date, our conclusion is that the return on the investment is not sensitive to the precise expenditure profile.

The appraisal is complicated by the replacement of one terminal by another. The existing terminal building was opened in 1972 and, on the basis of BAA's

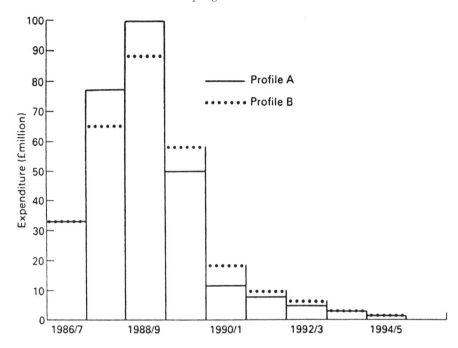

Figure 8.1 Alternative profiles of expenditure

standard accounting life for terminal of 25 years, will be life-expired by 1997. If the economic life of the existing terminal is realistically measured also at 25 years then the implication is that BAA would need to invest in some new facilities simply to serve existing traffic after 1997. Between 1991 and 1997, therefore, the financial benefits of the replacement of the existing facilities by the new terminal comprise:

- lower operating costs per unit of throughput on traffic that would otherwise have been handled by the existing facilities;
- incremental net revenue on traffic additional to that which would have been handled with the existing facilities; a profile of the phased build-up of traffic is shown in Figure 8.2.

From 1997 onwards the financial benefits of the new terminal comprise the net revenue earned from the total traffic throughput of 8 million passengers per annum. Clearly if the economic life of the existing facilities exceeds the accounting life of twenty-five years and they could continue to be used after 1997, then this calculation will tend to overstate the benefits of the investment.

From BAA material on the comparable northern terminal at Gatwick and Heathrow's Terminal 4, we assess the operating costs of the new Stansted terminal to be about £2.30 per passenger (at 1985 prices). This is probably about half the unit costs of the existing terminal when that is operating at capacity.

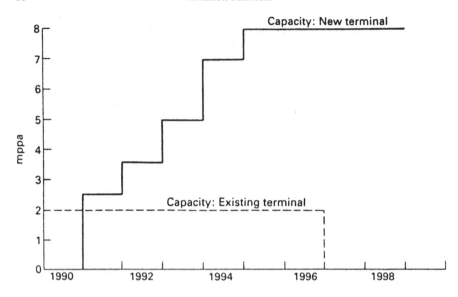

Figure 8.2   Terminal capacities and assumed traffic growth

The more important assumption concerns future net income. In 1984/85 gross income per passenger at Heathrow and Gatwick was £8.04 and £6.06 respectively (revalued to 1985 prices). This difference reflects lower income from traffic charges (aircraft landing and parking charges etc.) and lower sales per head of duty-free/tax-free goods at Gatwick. Stansted's gross income per passenger was £7.11; this was inflated by large duty-free sales (primarily to Scandinavian tourists). BAA's forward planning for Stansted has proceeded on the basis that, in the 1990s, it will emulate the present-day Gatwick and especially Gatwick's emphasis on whole-plane charter traffic.

We think, therefore, that Gatwick's gross income figure is a more reliable indicator of what one might expect at Stansted during the next decade if existing policies are maintained.

To calculate the net benefit stream, we have allowed for a net income of £3.76 per passenger (i.e, a gross income of £6.06 minus operating costs of £2.30 per passenger) for 8 mppa from 1997 onwards. In 1995 and 1996 we have allowed for an increment of 6 mppa, plus the saving in operating costs over the 2 mppa who would otherwise have used the existing terminal until 1997. Between 1991 and 1995 we have again allowed for the operating cost saving for 2 mppa, plus incremental revenue based on a phased build-up of traffic (as shown in Figure 8.2). On this basis the discounted net benefits sum to £201 mn (at 1985 prices), well short of the present value of the capital expenditures.

For the investment to break even on the basis of the various assumptions we have adopted, the net income per passenger has to increase to £4.48 which implies a gross income of £6.78 (see Table 8.1). The increase in real net income is almost 20 per cent.

Table 8.1   The relationship between gross income per passenger and net benefits

Gross income per passenger (£)	6.06	6.33	6.78	7.29	7.58
Discounted net benefits[1] (£mn)	201.00	214.50	236.60	261.60	275.70

1   As defined in the text.

It seems unlikely that BAA will be able to obtain from the discretionary spending on duty-free goods and other commercial services all the extra revenue required. In fact, we would expect commercial revenues at Stansted to decline substantially from present levels as the passenger mix becomes more varied and less dependent on Scandinavian traffic.

On the other hand, a substantial increase in traffic charges would highlight some awkward issues. BAA traditionally has set its charges at Gatwick below those at Heathrow to encourage traffic to 'spill over'. If the same policy is now adopted vis-a-vis Stansted, but Stansted's charges rise to cover costs, then a general increase in charges must occur. This will prove to be a difficult policy to adopt in view of the alacrity with which US airlines, in particular, challenge such increases.

There is, in addition, an implication for the traffic forecasts on which the expansion of Stansted is based. Experience has shown that forecasts prepared during the last 20 years have proved optimistic, in many instances wildly optimistic. This optimism is at least partially allowed for in the adoption of the 7 per cent rate of discount. But it needs to be recognised that any increase in the level of real charges will have a depressing effect upon traffic; more so if charges at Stansted are raised relative to those elsewhere.

Conclusions

In the above circumstances, privatisation of BAA (now likely in the spring of 1987) might have been expected to lead to a reappraisal of both the scale and pace of the Stansted project. But it now appears unlikely that such a reappraisal will take place. Early in June, the government announced that the shareholding of any individual, or of parties acting in concert, was to be restricted to 15 per cent of the total issued share capital and that the Secretary of State was to retain a special share with rights of veto.

This will weaken capital market constraints and increase managerial discretion to pursue non-profit-maximising strategies. The ability of managers to increase their own benefits at the expense of shareholders is limited, *inter alia*, by the threat of take-over from those 'raiders' identifying opportunities for improving profits. The government, by precluding such a possibility, has made it unlikely that the capital markets will be able to operate in this case with their usual efficiency in restraining over-ambitious investment.

There are further implications. If the capital market is well-informed and takes a less sanguine view of Stansted's expansion, as we have done, we would expect it to discount the value of BAA on privatisation. The proceeds from flotation will be reduced and the investment will not be, as Sir Norman Payne (Chairman of BAA) claimed recently, a case of an investment carried out at no cost to the taxpayer. Furthermore the early investment in additional capacity will place the privatised BAA in a strong position to compete for traffic with other airports.

The decision to allow the Stansted investment to go ahead immediately may thus have a number of adverse consequences. It appears especially curious that a government committed to the introduction of market forces into areas of decision-taking previously dominated by the nationalised industries and government departments should allow the market's judgement to be pre-empted in this way. It is also curious that, through the limitations placed on BAA's ownership structure, it should render the operation of market incentives after privatisation largely ineffective.

Chapter 9

Testing the Regulatory Model[1]

Introduction

The decision in the December 2003 White Paper, *The Future of Air Transport*, to support the expansion of London-Stansted airport was controversial, as was a similar expansion decision made in 1985. Much of the controversy has focused on planning and environmental issues, but both these decisions have had major implications for the economic regulation of airports; in this context, expansion of Stansted tested, and continues to test, the regulatory model. In particular, its expansion is relevant to the debate concerning investment incentives inherent in the RPI-X approach to economic regulation and whether the UK style of setting maximum prices (price caps) encourages the 'sweating of assets' at the expense of new investment. Stansted's expansion also has a wider significance: it suggests a willingness of the government and the competition authorities to accept the leveraging of market power in pursuit of perceived public-interest goals; it provides an insight into the behaviour of economic agents when capital market disciplines are mute; and it illustrates some unintended consequences that can follow from market intervention.

The structure of the paper is as follows. First, it outlines the policies for the restructuring, partial privatisation and economic regulation of the airport industry set out in the 1985 White Paper on airports policy [...]. The White Paper was accompanied by proposals for a substantial expansion of Stansted, and parliamentary undertakings were given regarding its financing; these are outlined. At the time, some commentators expressed doubts that the project on the scale proposed was financially viable, a view borne out subsequently by the airport's financial performance [...]. The resulting pricing policies led to a formal complaint of market distortion; the regulator's decision is reviewed [...]. An important feature of economic regulation applied to the London airports was that the allowable rate of return was based on their combined assets.

In practice, this led to the leveraging of Heathrow's market power in order to compensate for Stansted's poor return [...]. The motives and the regulatory risks that BAA, the airports company, faces are then considered [...], prior to noting that its risk profile has changed significantly as a result of the regulator's 2003 decision to base the price caps on the asset base of each airport [...]. This decision preceded publication of the 2003 White Paper, which announced the

1   First published in *Fiscal Studies*, 2004 and republished here by kind permission of the Institute for Fiscal Studies.

government's decision to support the development of a second runway at Stansted [...]. BAA argues that this latter project is not financially viable on a stand-alone basis in the time frame envisaged by the government, but is justified by public-interest issues; these are subject to detailed consideration [...]. The penultimate section considers the investment incentives in price-cap regulation in the light of the Stansted proposals [...]. The paper concludes by reflecting on the positions adopted by various stakeholders and by suggesting that a formal price cap for Stansted might no longer be appropriate.

The 1986 White Paper on Airports Policy

The current structure and ownership of the UK airport industry reflect in large measure the policies set out in the 1985 White Paper on airports policy (Secretary of State for Transport 1985), precursor to the 1986 Airports Act. The White Paper set out the government's views on the structure, ownership and regulation of UK airports in general and the future development of London's airports in particular. It stated that airport policy should be directed to making 'the best use of existing facilities and providing new capacity only when this is economically justified' (para. 3.1) and that 'it also continues to be the government's policy that air transport facilities should not in general be subsidised by the tax payer or the rate payer. Airports ... should normally operate as commercial undertakings' (para. 3.2). In the event and as we shall see, this commercial objective later proved difficult to reconcile with the government's concurrent decision that substantial, additional capacity in the south-east was to be met by the development of Stansted. First, however, I will set out the background to the White Paper and some of the policy details contained in it or in the ensuing legislation. Specifically, I consider the structure and ownership of airports with particular reference to the privatisation of the British Airports Authority; the framework of economic regulation established under the 1986 Airports Act; and the proposals to expand Stansted airport on a large scale. All three matters are interrelated in the ensuing argument.

Structure and Ownership of UK Airports

At the time of the White Paper, nearly all large airports were in the public sector and were owned either by the British Airports Authority or by various local governments. The Authority owned seven airports, four in Scotland and three – Heathrow, Gatwick and Stansted – in the London region; the only significant airport in the southeast it did not own was Luton. As a consequence, it dominated the national air transport market, accounting for 80 per cent of international passengers handled at all UK airports. The largest local authority airport was Manchester, with 6.0 million passengers per annum and a 10 per cent market share. Of the London airports, Heathrow was pre-eminent, with over 30 million annual passengers and a focus on premium long-haul international traffic. Gatwick, the UK's second-busiest airport, with nearly 15 million annual passengers, focused on charter and short-haul traffic (both the more price-elastic parts of the market), whilst Luton (1.6 million annual passengers) and little-used Stansted (0.5 million annual passengers) were dominated by charter carriers.[2]

The 1985 White Paper announced the government's intention that as many as possible of Britain's airports should become private sector companies. The British Airports Authority was to be privatised, with each of its seven airports organised as a separate company under a single holding company. It was not proposed to enforce the sale of local authority airports to the private sector, but all major airports were to be set up as airport companies to improve financial disciplines and to facilitate the transfer of ownership. The British Airports Authority was subsequently privatised (by flotation of shares) in 1987 and thus became BAA plc.[3]

One controversial aspect of these proposals was the intention to privatise as a single group all seven airports then controlled by the British Airports Authority. This had had been strongly argued for by the incumbent management, and at that time it was the practice to privatise the state utility industries without restructuring; BT and British Gas had already been privatised in this way. Not until later in the 1980s was greater emphasis placed during privatisation on facilitating competition by distinguishing the core natural monopoly elements of a utility enterprise and setting up the rest as separate potentially competing companies. Nevertheless, arguments for dividing the British Airports Authority were advanced by a number of parties at the time of the White Paper, particularly by the local authority airports led by Manchester. The government did not dismiss out of hand such possibilities but it concluded that competition between the airports was possible but limited, especially where Heathrow was concerned. On the other hand, dividing the ownership of the Authority's airports would, it was thought, have undermined an important plank of airports policy – namely, the Traffic Distribution Rules (TDRs).

The TDRs were a policy for intervening in the market, first introduced at Heathrow in 1977 when it was considered that problems caused by a shortfall of capacity at an existing airport could be tackled by directing traffic to another nearby airport. It was envisaged that further intervention would become necessary and that having an integrated airport system under common ownership would both facilitate the administration of such a policy and minimise its commercial

---

2   Since the mid-1980s, Heathrow, Gatwick and Stansted's combined share of the national market has declined a little but the three airports still retain over 90 per cent of traffic in London and the south-east. At Luton and Stansted, low-cost carriers have replaced much of the charter traffic and, for reasons set out below, Stansted has grown much more rapidly than Luton. Manchester still retains a market share of about 10 per cent. Passengers using low-cost carriers are particularly price sensitive.

3   The current ownership picture of UK airports is complex. In addition to BAA, some previous local authority airports ... are now wholly within the private sector whilst others have introduced private capital. There are two airports that were previously in the private sector but have now been taken over by the local authority-controlled Manchester Airport Group.

impact on any individual airport. Privatising the Authority's London airports as a group was consistent with this policy. In the event, in the early 1990s, the government decided to set aside the most important TDRs, thus removing the original *raison d'être* for having Heathrow, Gatwick and Stansted in integrated ownership.

Economic Regulation

The presumption at the time of the 1986 Airports Act was that major airports, like other public utilities, had significant market power and therefore should be subject to economic regulation. Although the Act brought all airports with an annual turnover exceeding £1.0 mn within the scope of economic regulation, more stringent regulation was reserved for those 'designated' by the Secretary of State under S.40 of the Airports Act. The airports designated following the 1986 Airports Act were Manchester, Heathrow, Gatwick and Stansted.

The regulatory process applied to these airports after they were designated mirrored that applied to the already privatised BT and British Gas, although there were some differences – notably, the important role given to the general competition body in advising the industry-specific regulator, the Civil Aviation Authority (CAA). The focus was on the so-called RPI-X approach to constraining the growth in prices: the increase in average charges is limited to the growth in the retail price index (RPI) minus a factor, X. The latter is set to allow for a rate of return consistent with the cost of capital. The return is calculated taking into account forecasts of: air traffic; total revenues (including substantial revenues from commercial activities such as airport retailing, known as the single till approach); net operating expenditure after taking into account feasible improvements in efficiency; depreciation of existing capital assets; and the proposed capital expenditure programme of the company. The last two components are the constituents of the regulatory asset base (RAB). It is intended that the company should earn an appropriate return on its RAB, so much time and effort are spent calculating the firm-specific cost of capital (using the Capital Asset Pricing Model). Subject to accurate forecasts in the round and to an accurate assessment of the cost of capital, the company is guaranteed a satisfactory return on its asset base, and this in turn is reflected, through X, in the prices it can charge.

Account is also taken of whether the projected cash flow is sufficient to sustain the investment programme and whether certain financial criteria, important to the credit rating of the company, are likely to be met. These factors are considered in a forward-looking manner and particular emphasis is placed on a forthcoming period, which is usually five years. Every five years, the process is repeated and X is reset. The approach is considered to provide the regulated company with incentives to seek out operating efficiencies because further gains not anticipated when setting X can be retained to enhance earnings per share.

Because of this incentive mechanism, the RPI-X model has been considered by many to be superior to rate-base or rate-of-return regulation. Rate-of-return regulation ties the allowed profits of the firm to its use of capital. This provides the firm with an incentive to substitute capital for labour and, with uncertainty present, could also lead to gold-plated investment and, in addition to the excessive capacity accompanying over-capitalisation, to excessive service quality (Averch and Johnson 1962). But RPI-X has also been criticised. Because it is said to focus on short-term efficiencies, the approach is argued to encourage 'asset sweating' and to give inadequate incentives to invest. Other critics have pointed out that, particularly with the need to reset X periodically, the RPI-X approach (with its emphasis on cost of capital and the RAB) has converged in practice to rate-of-return regulation (Stern 2003). This has led to the description of the UK approach as rate-of-return regulation with a regulatory lag (Armstrong, Cowan and Vickers 1994, 172; Giulietti and Waddams Price 2006).[4] It is an issue that I will return to after examining the behaviour of BAA in relation to Stansted.

The Expansion of Stansted

On the same day that the White Paper was published in June 1985, the government issued its decision letters following long public inquiries into proposals for the development of Stansted and Heathrow airports. The letters on the planning applications approved that for Stansted (with conditions) but turned down that for Heathrow. Stansted was a former wartime bomber base that had been incorporated into the British Airports Authority when the nationalised company was formed in 1966. The expansion project was strongly supported by the Authority's management, led by Sir Norman Payne (whose personal aim at Stansted was to create the 'best airport in the world').

The intention to expand Stansted on a substantial scale and to privatise it within the same company as Heathrow and Gatwick led the local authority airports, of which nearby Luton was one, to express concern that this would lead to its cross-subsidisation to their disadvantage. Much of the ensuing debate in Parliament focused on this issue. Nicholas Ridley, the then Secretary of State for Transport, tried to allay these fears on a number of occasions during the debate on the White Paper (*Hansard*, 5 June 1985):

> I can confirm that there will be no subsidy to Stansted. The only investment will be of commercially raised money paying commercial rates of interest. That will cause landing charges to rise at Stansted to levels commensurate with other airports (column 315);

> I confirm that Stansted has benefited financially from the profits of other BAA airports. This should not be the case in future under our proposals (column 317);

> I have tried through the subsidy assurance ... to make this development happen only if it is necessary and economic (column 321);

---

4   Sibley (2000) also points out that rate-of-return regulation, as now applied, typically allows the firm to keep economic profits earned between rate reviews, and therefore it is suggested that there is probably little to choose between the two regulatory models. For a good review of the earlier literature on rate-of-return regulation, see Train (1991).

As for cross subsidy and fair competition, putting each airport into a separate public limited company means that any transfer of funds will be transparent. As I said, we shall make sure that the transfer of funds takes place only at commercial rates of interest. That in its self will cut out cross subsidy (column 325);

I have given every assurance this afternoon that there will be no opportunity for a cross subsidy to Stansted to the detriment of other competing airports. Market forces will prevail to the extent that it will be for the proprietors of airports, whether Stansted, Manchester or any other to decide whether investments are made at commercial rates of interest (column 329).

The government's intention, therefore, was that Stansted should be financed directly by the capital markets or indirectly by inter-company loans made at commercial rates of interest. The development was also to reflect market forces and to be viable. As we shall see, the expansion of Stansted did lead to cross-subsidies that were to the detriment of other competing airports, it has not proven to be a viable expansion by normal commercial standards and average charges have still to rise to levels commensurate with those of other airports.

Financial Forecasts and Performance

Financial Forecasts

At the time of the Commons debate, research at the Institute for Fiscal Studies (IFS) was examining in detail the commercial case for the proposed expansion of Stansted. In the 1985 White Paper, the government had stated that it would expect BAA to show that the investment would earn an acceptable rate of return. The researchers were sceptical that this would be the case (Hill, Starkie and Thompson 1985; Starkie and Thompson 1985, 1986a and 1986b).

The proposal was to construct initially a terminal capable of handling up to 8 million passengers per annum, together with the associated infrastructure (aircraft aprons, taxiways and access roads) necessary to support this volume of traffic. The new terminal was planned to have about four times the capacity of the then under-utilised terminal; the latter was to be replaced, giving a net addition in the first phase of 6 million passengers per annum. At the time, this was broadly equivalent to the number of passengers using Manchester, the country's third-largest airport.

The analysis took the traffic forecasts set out in the 1985 White Paper together with the cost estimates that had been presented to Parliament. Assumptions were made concerning the precise phasing of construction expenditures, the running costs of the proposed terminal (these were based on the figures for Gatwick), the expected economic life of the investment and, importantly, the net income to be expected from the new facilities (this too was based on Gatwick figures).

The conclusion of the research was that net present revenues would fall well short of the net present cost of the investment. For the investment to break even, net income per passenger would have had to increase in real terms by almost 20 per cent, at which level it would be well above the comparable Gatwick figure. The bottom line was that the scale of the proposed investment was considered to be suspect and not driven by market forces, a view supported by Stansted's subsequent financial performance.

Financial Performance

At the time of the mid-1980s debate on airports policy, Stansted was making a large loss, before capital charges, equivalent to 40 per cent of turnover. It had recorded a loss on operating account for the previous 14 years. Not since 1970–71 had it made an operating surplus, and this was at a time when it had less traffic than in 1984–85.

There were a number of probable reasons for this poor performance. First, Stansted's terminal capacity was at the time well in excess of that required for its traffic. The 1985 White Paper showed that it had more spare terminal capacity than any other UK airport bar one (Liverpool) so that capital charges and operational overheads were disproportionately high. Second, there were indications that the pre-privatised British Airports Authority was providing an excessive level of service in relation to traffic levels. For example, the airport was open 24 hours in spite of a trivial amount of night-time traffic. Third, revenue yields per passenger were also comparatively low and had been declining over the longer term.

By 1990–91, shortly before completion of the major expansion in terminal facilities, annual passenger traffic had grown to 1.1 million, a little more than half the notional capacity of the old terminal, and the operating loss by now exceeded 100 per cent of turnover. Four years later, in 1994–95, Stansted's financial performance had improved (see appendix 5.9 of Monopolies and Mergers Commission (1996)) but the loss was still considerable, at approximately one-third of turnover. Not until 1997–98 (see appendix 4.7 of Monopolies and Mergers Commission (1996)) did Stansted make an operating profit, at which point it had been making losses consistently for over a quarter of a century.

Operating profits continued to increase, slowly, into the twenty-first century, but it was evident from the Competition Commission's 2002 review of BAA's London airport companies that the airport still lacked market power; it was estimated that Stansted's revenue per passenger from airport charges was £2.99 net of adjustments (compared with an estimated £4.08 at Gatwick). Overall, the airport continued to earn a return below its cost of capital and it was expected to continue to do so until 2008, when it was assumed by the Commission that Stansted's net yield would increase gradually towards Gatwick's projected yield (Competition Commission 2002, para. 2.310). But, even if this assumption proved accurate, the equivalence of the investment return and the cost of capital would come almost 20 years after large-scale expansion of Stansted commenced.

Thus, Stansted's financial performance has been consistently poor over several decades and the airport still makes a return less than its cost of capital. In normal commercial circumstances, one might have expected closure of the facility or restructuring as a smaller but financially sustainable operation. Why BAA has been willing for Stansted to sustain such losses over such a period of time is a subject to which I will return. But first it is worth noting one adverse consequence of Stansted's poor financial performance.

Distortion of Competition

BAA's three London airports were kept together at privatisation purportedly to avoid the adverse consequences for any one airport of a change in the TDRs. This policy stance overlooked the position of Luton, sandwiched between Stansted and Heathrow. Another argument advanced – that the opportunities for competition between airports were limited – was an equally anomalous view (Starkie, 1994). It sat rather awkwardly with S.41(3)(c) of the 1986 Airports Act. This section gives the CAA, as the regulator, the power to impose conditions when it is judged that an airport operator has fixed charges that are insufficient to cover the cost of a service or facility and those charges materially harm the business of another airport operator. But for pricing at one airport to be harmful to another airport, there has to be a prospect for serious competition, a notion that the White Paper rejected when it argued against the breaking-up of BAA.

In June 1993, Luton airport complained to the CAA that Stansted was in fact pursuing a course of conduct as described in S.41(3)(c). In essence, the complaint was that Stansted was levying charges that were too low and that these low charges were materially harming Luton because they had led to the transfer of traffic from Luton to Stansted. The specific example was given of the transfer of Ryanair services.

In response, the airport company accepted that it was not making a return on capital employed or even a profit on book cost, but argued that to be expected to do so in the early days of a major development would be unreasonable. As Stansted's revenues exceeded those costs that would be avoided by closing the airport down, BAA maintained that there could be no finding of unreasonable or anti-competitive behaviour. Closing the airport or putting it on a care-and-maintenance basis would be most unlikely to outweigh the loss of revenue. Stansted was not subsidised, BAA argued, but borrowed from the Group on fully commercial terms.

The CAA formally responded to the complaint (Civil Aviation Authority, 1993). It accepted as 'strictly true' BAA's denial that it subsidised Stansted but, given the state of Stansted's balance sheet, it was thought 'improbable' that the borrowings could have been obtained except from fellow members of the Group (para. 49). It also said that it would be inconsistent with the CAA's earlier decision to treat BAA's three London airports as a system if it did not 'allow at least a measure of cross-subsidy' (para. 43). It nevertheless conceded that the scale and timing of the expansion of Stansted had been premature, and it was the excess capacity that was at the root of the problem. In the circumstances, the CAA accepted that Stansted had little choice but to engage in heavy promotion on a scale that might not otherwise have been prudent or justifiable. Closing the airport or placing it on a care-and-maintenance basis would have done considerable damage to the airport's future prospects. The CAA did recognise that Stansted's *net* charges (i.e. after discounts) to airlines clearly failed the sufficiency test and it accepted Luton's contention that Stansted's pricing policy was causing it material harm. In these circumstances, the CAA found that the S.41(3)(c) tests were met.

However, the CAA pointed out that the Statute provides for a remedy only if the CAA thinks fit. This appeared to be the crucial point. The CAA was unable to come up with a realistic remedy, given that closure or contraction was not considered feasible. The best that could be achieved in the circumstances was to follow a policy of loss minimisation. It added:

> Luton has been materially harmed but that harm stems from the Government's decision in 1985 that the next significant tranche of airport capacity to serve the South East should be at Stansted ... For the Authority to find that the policies Stansted has pursued ... should be inhibited or reversed because of their effect on Luton would be tantamount to saying that the decision to develop Stansted was wrong. (para. 61)

The CAA went on to conclude that no remedy was appropriate.

The CAA obviously had to make a very difficult decision. Faced with circumstances where infrastructure costs were sunk and operating costs were low, it probably made the correct economic decision overall, but in reaching the conclusion that no remedy was appropriate, it is open to criticism that it took too much on trust from BAA. For example, the CAA did not appear to consider whether closure of *some* facilities on a temporary basis might have been worthwhile. The CAA could also be criticised for failing to challenge BAA as to whether contracts with each individual airline were priced at least to recover marginal costs. The economic literature suggests that in the circumstances where an incumbent's average price level is regulated, as is the case here, the incumbent has an incentive to price-discriminate and to price below marginal cost (Armstrong and Vickers 1993). There are strong indications that this is what was happening at Stansted.[5]

How Does BAA Balance the Books?

Until the latter half of the 1990s, Stansted had been returning operating losses for about 25 years. Why, therefore, did BAA, a major public limited company, persevere with the ambitious investment plans inherited from the British Airports Authority? Furthermore, with Stansted still not expected to produce annual returns equivalent to its cost of capital until 2008, why has BAA not taken drastic steps to curb losses at the Stansted subsidiary? Part of the answer to these questions is to be found in the willingness of the government to designate Stansted

---

5   In 1996, three years after the CAA's decision, *net* revenues from traffic charges at Stansted did not cover the direct or volume-related costs of providing for Ryanair, whose move from Luton had instigated the original complaint (Monopolies and Mergers Commission 1996).

and thus bring it within the scope of S.40 of the 1986 Airports Act, and part of the answer is to be found in the way that the industry regulator, the CAA, has applied economic regulation.

Designation of Stansted was, on the face of it, a peculiar decision. To suggest that, at the time of designation in 1987, Stansted had significant market power was patently absurd. Even in 1996, after a significant growth of traffic, the Monopolies and Mergers Commission was to note that 'As regards STAL [Stansted], the CAA suggested to us that there seemed to be little monopolistic power and very little reason to apply a maximum to charges which would be likely to be binding in the immediate future' (Monopolies and Mergers Commission 1996). But, given Stansted's weak market position (its relative inaccessibility and concentration on price-elastic market segments), it was important for BAA that it was designated if the major plans for expansion were to go ahead without detriment to shareholder interests; the government obliged.

Designation was not by itself, however, a sufficient condition for securing shareholder interests; it was also necessary to link the Stansted expansion with the pricing power of Heathrow; the latter is considerable because of high levels of demand for access from international scheduled carriers relative to the runway and terminal capacity available (expansion of which is subject to severe planning and environmental constraints). The link between Stansted and Heathrow was achieved by a particular treatment of BAA's assets in the regulatory process, which was to calculate the allowable rate of return on the *combined* assets of Heathrow, Gatwick and Stansted – what is referred to as a system-wide RAB. Consequently, where any one airport is unable to earn its cost of capital, prices are adjusted elsewhere in the system so that the system as a whole earns its cost of capital (see Competition Commission (2002, para. 10.36)).

Thus, the airport system approach to the RAB ensures that the financial deficits of Stansted are recouped through higher charges at Heathrow. BAA is prevented by the regulatory price cap from charging market-clearing prices at Heathrow and capturing the scarcity rents associated with its capacity constraints, but, in effect, it has been permitted to set prices that sufficiently leverage its market power to ensure that, across its three London airports when taken together, the books are balanced. As a consequence, there has been no financial penalty from excessive expansion of Stansted; the combination of an allowable return reflecting the cost of capital, a system RAB and considerable market power in at least part of the system leads to a satisfactory outcome for BAA.

Finally on this issue, it should be noted that the taxpayer has also lost out as a consequence of Stansted's overexpansion and generally poor financial performance. Stansted has been able to sell its tax losses to other Group companies (albeit at the appropriate rate so that there has been no subsidy hidden as part of the transaction). But the other Group companies have been able to offset these losses against their profits, thus reducing, overall, the Group's tax liabilities. This practice of trading tax losses within a group of companies is a legitimate practice, but, because of the losses caused by overexpansion at Stansted, general tax revenues are less than they would otherwise have been. It is also possible that the sale of these tax losses constitutes the icing on the BAA cake; it is not clear that adjustments were made for this tax benefit when setting the regulatory price caps (which is done on an after-tax basis).

Motivation and Risk

There is now an extensive economic literature on the principal-agent problem – for example, the problem that owners of enterprise capital have in ensuring that managers pursue objectives in their interests.[6] The private sector attempts to resolve the problem by giving management financial incentives that are compatible with shareholder interests. In addition, discipline also comes from the threat that underperforming companies are liable to be taken over, putting management jobs at risk.

Arguably, this latter discipline is weaker the greater the capitalisation of the company. Consequently, privatising the former British Airports Authority as a whole would, in normal circumstances, have reduced the threat of a subsequent takeover. But in this particular case, a restriction was written into BAA's Articles of Association at the time of privatisation, limiting any single shareholder to 15 per cent of the equity. As a result of a recent ruling in the European Court concerning the free movement of capital, the restriction is to be removed, subject to a successful resolution, at the 2004 AGM, but thus far, because of it, it has been impossible to subject BAA to a hostile bid. Possibly reflecting these circumstances, it has been the major UK airlines in particular that have lobbied for the government to intervene to break up BAA. The system RAB approach does, however, make this a more difficult proposition, which is likely to be an added reason why the approach has found favour with BAA.

Given these circumstances, has BAA's behaviour with respect to Stansted been consistent with maximising shareholder value? There are indications that BAA's management has been pursuing a policy of maximising market share by building excess capacity and then pricing in the Stansted market at (or, at peak times, below) short-run marginal costs[7] and recouping total costs, including the investment costs, from other airports where demand is less elastic. Nevertheless, in the peculiar circumstances that Stansted has found itself in, it is possible that maximising market share through investing in excess capacity has not been inconsistent with maximising shareholder value. Incorporation of Stansted in a system-wide RAB and the fact that BAA is allowed its cost of capital across the three airports combined mean, in effect, that any financial downside from pursuing a market-share strategy has been compensated for by increased charges and the extraction of rents at Heathrow.

It is, however, a strategy not without risks attached, depending, as it has done, on regulatory acceptance of a system approach. Absent the system approach, the

---

6   Vickers and Yarrow (1988) provide a good review of the general problem

7   There are now well-defined peak periods at Stansted during the course of the day. Peak demand normally drives the investment case and therefore peak pricing is a way of testing whether and when additional capacity is justified.

lack of a satisfactory return on Stansted's capital would have forced operating efficiencies, pricing adjustments and a probable write-off of capital. Consequently, the regulatory review leading to the fixing of the price cap for BAA's London airports for the period 2003–08 is of particular significance.

The 2003 Price-Cap Decision

In its review of the 2003–08 price cap (Civil Aviation Authority, 2002a), the CAA indicated at an early stage that it proposed to set aside the system approach and consider each London airport separately. It considered the case for stand-alone price caps to be very strong for two reasons. First, the demand and capacity conditions were different at each airport, with each airport largely serving separate markets. It therefore considered that a focus on airport-specific caps allowed for a better tailoring of incentives to the needs of each individual airport. Second, it also considered that the use of an airport's own asset base to fix its price cap diminished the ability of the airport operator to subsidise 'premature or gold-plated investment'.

The Competition Commission in its October 2002 recommendations to the CAA suggested that the system RAB approach should continue. It did not accept that because airlines are not prepared to switch airports, investment at Stansted does not reduce excess demand at Heathrow.[8] Under a separate airport approach, if Stansted were unable to price up to a cap that reflected its own cost of capital, the return of the three airports as a whole would fall short of BAA's cost of capital.[9] This would be inconsistent with the previous basis of regulation and would put at risk BAA's ability to finance its investment (para. 2.51). The Competition Commission also said that it saw no evidence of undue investment at Stansted and no evidence that investment was predictably excessive nor, therefore, that a system RAB was driving such an outcome (para. 2.54).[10] The Commission also shared the concern of the Department for Transport that a stand-alone approach could inhibit the financing of additional runways at Stansted, the 'need' for which may rise because of constraints on runway developments elsewhere (para. 2.56).

The CAA in its November 2002 *Proposals for Consultations* rejected this view and restated its intention to adopt a regulatory approach for the future based on

---

8   It also quoted BAA as saying that airlines mainly operated from Stansted because they were unable to operate from Heathrow (para. 2.45). It is difficult to conceive of the low-cost airlines that dominate Stansted moving to Heathrow if room were made available; the characteristics of Heathrow make it unsuitable for the low-cost carrier business model with the emphasis of the latter on fast turnrounds.

9   Elsewhere in its report, the Commission anticipated that Stansted would be meeting its cost of capital by the end of the current quinquennium.

10   This comment should be set alongside those in paragraph 1.14, which suggests that the Commission was reluctant to get to grips with BAA's capital expenditure programme. Its consultants were also critical of some of BAA's procurement processes (see paras 9.66–9.75).

the stand-alone regulation of each airport. This proposal was supported during the consultation phase by a number of airlines, particularly those operating from Heathrow. Airlines operating from Stansted took an opposing view. BAA, as might have been expected, argued strongly against the proposal, suggesting that the change of policy was premature in pre-empting the government's forthcoming White Paper; the Department for Transport made the same point.

In its *Decision* in February 2003, the CAA confirmed its position that regulation was to be based on a stand-alone approach with the price cap for each airport set in accordance with its own RAB. This decision did not affect the price caps during the period 2003–08 because these were fixed taking into account the expected return on capital over the forthcoming decade and it was anticipated that Stansted would, finally, produce an adequate return over this time horizon. Nevertheless, the CAA left the door open for a return to the *status quo ante*, but confirmed its earlier proposal that it would expect 'compelling evidence to demonstrate that users in aggregate were genuinely better off as a result' and, no doubt with the Luton case in mind, 'that the impact of [a system-wide RAB] was not unduly distortionary or discriminatory as regards other airports in the southeast' (para. 3.24).

From BAA's point of view, this change of policy on the part of the regulator is of critical importance. It means that, at least for the time being, BAA cannot rely on securing a return on further investment at Stansted by being allowed to leverage its Heathrow market power. And in the event that the CAA can be persuaded to adopt a system approach in the future, the current decision will have increased perceptions of the regulatory risks involved.

The 2003 White Paper

The purpose of the White Paper, *The Future of Air Transport*, published in December 2003, was to set out the strategic framework for the development of airport capacity in the UK over a period of 30 years.[11] Within this framework, airport owners were expected to bring forward proposals to be considered through the planning system in the normal way. The government saw its role as primarily one of enabler and regulator, given that the major airports are, in the main, operated within the private sector.

Publication of the White Paper was preceded by an extensive consultation process, the release of a number of technical papers by the Department for Transport, including regional studies of site options for airport expansion, and the subsequent release of the responses to the consultation. The BAA submission (BAA, 2003) reviewed the various site options and was of the view that 'one additional runway at Stansted would be financially viable, subject to the scale of the additional costs not calculated [in the preliminary studies], although the

---

11   In contrast to its 1985 predecessor White Paper, it did not contain any legislative proposals.

charges needed to remunerate the investment would need to be shared across users of the London system as a whole' (para. 7.51).

BAA's analysis had suggested that charges needed to remunerate the investment would have to increase above the 2003–04 level (for Stansted) in real terms by around 35 per cent if approached on a system basis and by around 120 per cent on a stand-alone basis. To more than double real charges was not considered feasible. Thus, an additional runway at Stansted was *not* viewed as commercially viable without cross-subsidy from other London airports (or, alternatively, without direct subsidy). BAA went on to argue in favour of cross-subsidy through the system approach invoking what might be thought of as public-interest arguments in doing so. It suggested that maintaining the system approach would continue to be justified on the grounds that all users would benefit from the provision of additional capacity in the southeast, irrespective of its location as a result of the reduced congestion, enhanced airline competition and lower air fares. There would be wider public interest benefits to developing airport infrastructure in a way that underpinned the economic health of London, the southeast and the UK as a whole.

As for the White Paper itself, and not withstanding BAA's conservative response on the commercial viability of further investment at Stansted during the consultation phase, its major pronouncement was that 'the Government now supports the development of a second runway at Stansted as the first new runway to be built in the South East' (para. 11.40). It went on to add that:

> The Government will not promote or pay for the development of Stansted. New airport capacity should be paid for by airport users. We look to the airport operator to take it forward in a way that is responsive to users, and to provide necessary funding. It is a responsibility of the regulator, the CAA, amongst its statutory duties, to encourage timely investment. The Government expects … Regulator and airport operator, to secure an appropriate framework to bring the development to fruition. It expects this process to be guided by the decisions in this White Paper, as well as by the regulator's duties towards users of airports, towards the operation of airports, and towards investment in new facilities at airports.

One could interpret this statement with its emphasis on the regulator's statutory duties as an indication that the government sees the system approach as the appropriate means for funding the further expansion of Stansted. The reference to statutory duty is a reference to Section 39 of the 1986 Airports Act that *inter alia* requires the regulator 'to encourage investment in new facilities at airports in time to satisfy anticipated demands by users of such airports'. But it does beg the question of what is timely in these circumstances and whether promoting an investment that is not financially sound would breach the Statute, bearing in mind that another clause in Section 39 places a duty on the regulator 'to promote efficient, economic and profitable operations of such airports'. If the government were to favour the former clause, then it would also appear to contradict its own appraisal framework set out in the consultation, which stated that commercial viability was 'a hurdle that must be passed' by developments on new or existing airport sites (Department of the Environment, Transport and the Regions 2000, para. 13.1).

The response of BAA to the White Paper was to announce that it would press ahead with the expansion of Stansted, that the process of seeking planning permission had begun and, interestingly, that it hoped to reduce significantly its earlier estimated costs of expansion (these estimates had formed the basis for the calculated increase in the level of charges required to remunerate the investment).[12] BAA warned, however, that by adopting a stand-alone approach to financing the expansion, there was considerable risk that timely provision would be prejudiced and that this might make government subsidy necessary, adding that without financial support from Government, the airport operator would have to wait until demand, and hence airlines' willingness to pay, for new capacity had risen to a level that would justify investment in the new runway. 'Such a delay in providing new capacity would not be in the interests of airlines, their passengers or the wider economy' (*Business Weekly*, 28 November 2003). Yet again, BAA was suggesting that there were public-interest arguments justifying either leveraging its market power at Heathrow (to cover the projected financial shortfall of an early expansion at Stansted) or the payment of a direct subsidy. In its response to the consultation, it had advanced two such arguments for an early addition to Stansted's capacity: reduced congestion elsewhere in the system and enhanced airline competition leading to lower airfares. It is to these two arguments that I now turn.

Public-Interest Arguments

Relief of Congestion Elsewhere in the Airport System

The first public-interest argument put forward by BAA to justify financial (cross-)subsidy of Stansted is that it would reduce charges at Stansted (albeit, in the case of cross-subsidy, at the expense of higher charges at Heathrow/Gatwick); this attracts traffic to Stansted that would otherwise use busier airports, and thus congestion costs are reduced at the latter. This positive externality is to be allowed for over and above any financial margin. Before examining the essence of the argument further, a few comments on the size and nature of airport congestion are called for.

First, it is usual to declare a capacity for the runway(s) at an airport, defined as the permitted maximum number of aircraft movements in a specified time period; this in turn constrains the scheduling of services. The declared capacity for runways might also reflect constraints elsewhere in the airport system, so that

12   This might suggest that removing the system RAB has already led to a more careful examination of investment proposals. It is worth noting that the cost of a terminal expansion opened at Stansted in 2002 was, on a square-metre basis, about double the estimates used by the Department for Transport's consultants in the airport options study leading up to the 2003 White Paper.

permitted runway use is used to balance the airport system as a whole. Defining capacity also involves specifying service quality. For example, the degree of average delay at Heathrow agreed as acceptable is ten minutes. These administrative rules reduce the amount of congestion in the system and therefore reduce the potential benefits of reduced congestion if flights switch between airports. Second, recent contributions to the theory of airport congestion have argued that the traditional approach overestimates the marginal costs involved (Brueckner 2002; Mayer and Sinai 2002). It does not take into account the fact that a particular flight might very well impose delay on other flights operated by the same airline and that the delay costs that an airline imposes upon itself are, in effect, internalised. Third, the economically efficient response to congestion is to adjust prices where the congestion occurs. Although use of the price system in the airport context is constrained by international treaties and obligations, there is a surcharge for the use of runways at Heathrow and Gatwick during busy times together with a reasonably well-functioning secondary market in slots. These measures also help to reduce the scale of the congestion externality as well as assisting with the efficient allocation of existing capacity.

Putting to one side these considerations (all of which suggest that the congestion problem at crowded airports with dominant users is perhaps not of great consequence), the nub of the BAA argument, although one that is implicit, is that the reduction in the congestion externality as a result of flights switching to Stansted will exceed the 'deadweight' loss as a result of the use of resources priced at less than their (marginal) value in alternative uses. The argument hinges therefore upon the extent to which airlines using Heathrow and Gatwick will be attracted to Stansted by reduced charges. The smaller is the degree of switching of flights from one airport to another relative to the overall increase in the demand for flights at the airport where the price is reduced (i.e. the smaller is the cross-price elasticity relative to the direct price elasticity), the less likely it is that the reduced congestion externality will exceed the deadweight loss.

In the case of Stansted, the indications are that the elasticity conditions necessary to secure a net welfare gain are most unlikely to occur. In spite of considerable and long-standing excess demand for the use of Heathrow at current levels of charge and in spite of low charges at Stansted, Heathrow operators have shown no inclination to transfer to Stansted. But, because of pent-up demand at Heathrow (with airlines seeking many more slots than are available), to make an appreciable difference to congestion at Heathrow, switching would have to be considerable. Switching between Gatwick and Stansted is more likely, given the current structure of the market, but currently congestion at Gatwick is not especially severe. On the other hand, the direct price elasticity for air travel from Stansted on low-cost carriers is thought to be high. Consequently, it seems most unlikely that subsidising charges at Stansted in order to relieve congestion at Heathrow or Gatwick will result in a net gain in welfare.

Increasing Competition in Air Services

BAA's second public-interest argument starts from the premiss (which again is implicit) that in order for there to be a welfare gain, airlines do not have to respond to pricing differentials by transferring flights between airports. Instead, the generous provision of capacity at Stansted together with low charges encourages competitive entry into the market for air services and thus lowers airfares at other airports. Although the subsidy that is necessary to secure this outcome results in some loss of welfare (the resource costs of expanding Stansted exceed the net benefit to Stansted passengers), passengers at congested airports where competitive entry is not possible also gain from the lower airfares at Stansted: economy fares on short-haul routes out of Heathrow, for example, will be reduced by the increased (subsidised) competition from Stansted. This gain is assumed to offset the uneconomic use of resources at Stansted, resulting in a net gain overall.

This argument perhaps has greater merit but it also retains some significant weaknesses. Although the price mechanism is not used at busy airports to clear the market for the use of runways and terminals by airlines, prices are still used to ration passenger numbers to available seat capacity. This is done by the airlines using yield management techniques: the balance of seats between different classes and the price of those seats are continually monitored and adjusted as the number of seat bookings on a particular flight increases towards the maximum number of seats available. The result is that both the seat load factor and the average price of the ticket (the yield) increase with demand, and where demand is high the airline is able to secure a margin on sales in excess of, and perhaps well in excess of, operating costs. This excess is a scarcity rent.

The CAA has argued that airlines operating out of Heathrow in particular gain a significant scarcity rent. During the recent review of the airport price caps, it undertook research that showed that revenues from flights to a number of destinations from Heathrow greatly exceeded those from similar flights from Gatwick. It estimated, for example, that a BA short-haul flight operating out of Gatwick would show an additional profit of £2 mn per year at Heathrow. This difference, referred to as the Heathrow premium, does not take account of the higher operating costs experienced at Heathrow (Bishop and Thompson 1992); therefore the net premium is likely to be less but probably remains substantial.

In relation to the proposition that (cross-)subsidising flights out of Stansted has the beneficial effect of reducing fares at congested airports such as Heathrow, the scarcity rent or fares premium is of significance. In so far as fares at Heathrow are reduced by competition from Stansted (and at the moment there are relatively few common destinations served from both Heathrow and Stansted), the likely effect will be to reduce the fares premium at Heathrow and thus the scarcity rents. However, and importantly, this is not an efficiency gain but a rent transfer; the scarcity rent will, in effect, have been transferred from Heathrow airlines (a reduction in producer surplus) to Heathrow passengers (an increase in consumer surplus): a situation that will continue so long as demand at Heathrow exceeds capacity.

This is not to argue that there could not be some efficiency gain as a result of subsidised fares at Stansted. Even in a more crowded future at Heathrow, there will be times, particularly in the winter season, when demand is less, so that there is the potential for selling more capacity; the ensuing increase in output would represent a welfare gain. But it is doubtful that this gain would be large enough to offset the net loss at Stansted (where the resource cost of the additional, subsidised output will exceed its value to Stansted passengers).

British Airways (BA), on the other hand, argues strongly that a fares premium at Heathrow, where it holds about 40 per cent of the slots, does not exist; it points out that if such a premium did exist, it would be reflected in excess profits and return on capital, which BA does not experience. If there were no scarcity rent at Heathrow, this would strengthen the argument that subsidy at Stansted could beneficially impact on fares at Heathrow (but output would still need to increase). However, it is difficult to accept, in the light of the CAA evidence and more particularly in view of BA's willingness to pay other airlines operating at Heathrow considerable sums for an exchange of their slots, that there is no Heathrow premium.[13]

A fares premium at Heathrow and the absence of excess profits for BA can be reconciled, however, by the fact that BA has an extensive network of routes, not only at Heathrow but also out of Gatwick and other regional centres such as Birmingham, Manchester and Glasgow. On the fringes of these networks, BA probably faces strong competition; some services will be marginal and others might well operate at a short-term loss, albeit with the prospect of more profitable operations in the future. Other routes might be marginal even if competition is absent because of low levels of demand. It is possible, therefore, that BA uses rents from Heathrow to sustain a larger network and a greater service frequency (particularly to feed traffic to long-haul services) than it would otherwise do in the absence of those rents. Consequently, if competition from subsidised routes at Stansted cut into the Heathrow fares premium, BA's more marginal services, particularly those out of regional airports, could be at risk. Other airlines might respond to cuts by BA by increasing their services at these airports, but the extent to which they are able to do so will depend upon their having a much lower cost base than BA in comparable circumstances (to compensate, in their case, for the absence of a Heathrow premium). A response from other airlines is also less likely if BA responds to pressure on fare premiums at Heathrow by cutting frequencies rather than routes. Similar arguments are likely to apply to BMI, the airline with the second-largest slot portfolio at Heathrow. Therefore, the likely extent and the direction of change in both prices and output across all the relevant networks are not at all clear.

In addition, account also needs to be taken of the likelihood that further Stansted subsidy would again lead to a serious distortion of competition between airports. It was accepted by the CAA that Luton Airport was materially harmed by the last major expansion at Stansted, and other airports in the Midlands and

---

13   Press reports in October 2003 suggested that BA had finalised the acquisition of two pairs of slots for which it paid £12 mn.

East Anglia were probably affected too (although to a lesser degree than Luton). At the time of Luton's complaint, it was argued that there was no suitable remedy available; the excessive capacity at Stansted had already been 'sunk' and therefore low prices were inevitable if available capacity was to be used efficiently. This argument does not apply in the current pre-commitment context. Moreover, although there might in the past have been benefits from bringing forward airline competition at some cost to competing airports by expanding Stansted prematurely, it is difficult to justify further premature investment on the same basis now that low-cost airline competition is in place.

Price-Cap Regulation and Investment Incentives

The Stansted expansion project clearly raises a number of issues regarding both the structure of the airports industry and particularly the regulatory framework that is applied to it. The UK approach to economic regulation operates by restraining average prices (in the airports case, average revenues) in a way that provides the regulated firm with incentives for efficiency because subsequent savings in expenditure, exceeding those anticipated at the time that the regulated price (revenue yield) is set, enhance company profitability. But, as I noted earlier, some of the recent literature has argued that this approach fails to give adequate incentives to invest because of problems of regulatory commitment, the longevity of sunk investment and the *ex-post* opportunism that this can give rise to, and the general focus of the approach on short-term efficiencies. On the other hand, others (for example, Littlechild 2003) have noted that pressures on regulators have led, in practice, to a steady convergence of the RPI-X approach with rate-of-return regulation, the latter generally regarded as encouraging overcapitalisation and excess capacity.

BAA's approach at Stansted appears to support the view that an RPI-X approach does not inhibit the regulated firm from investing in capacity. The company's apparent willingness, verging on zeal, to expand capacity at Stansted in advance of a level of demand that would justify the investment indicates a strong disposition to overcapitalise (as a firm subject to traditional rate-of-return regulation would do). Similar indications arise from BAA's reluctance, in spite of peakiness in the use of infrastructure, to use time-of-day pricing to test the strength of demand. But it is a predisposition that the RPI-X approach appears to do little to restrain. Because the regulator is setting a price, usually every five years, with reference to an allowable return on assets, the regulated firm with market power has an incentive to expand its asset base provided that the regulator allows for this in the prices it can charge.

The regulator can, of course, try to press down on such tendencies by scrutinising the proposed capital expenditure programme when resetting X. The regulator of the water utilities, for example, does this, but it has not been the practice in the airports industry. Generally speaking, the capital expenditure programme put forward by BAA has been accepted without detailed examination. During the recent periodic review, for example, the Competition Commission commented that 'we have not adjusted BAA's forecast for capital expenditure: even if there is scope for lower costs on some projects, there is in our view likely to be a demand for any cost savings to be spent on additional projects' (Competition Commission 2002, para. 1.14). The CAA has also been reluctant to challenge details of the capital expenditure programme, believing that this would result in too intrusive regulation (see, for example, Civil Aviation Authority 2002a, para. 3.27). Instead, it has sought to encourage the airport companies and the airlines to engage in a process of consultation and mutual agreement on an appropriate programme of works; but the airlines have viewed the process as one that is less than satisfactory (see, for example, Competition Commission 2002, para. 9.84).[14]

The consequence of this regulatory approach is that it has provided an opportunity for airport companies to exercise their market power indirectly through higher-than-necessary levels of investment in capacity and in costly gold-plated investment. In the specific case of Stansted, this process has been taken a step further: BAA has been allowed to leverage market power at Heathrow in order to finance an expanded asset base in the distinctly different Stansted market so as to provide services in the latter at (short-run) marginal cost. This has been at the expense of competing airports, particularly Luton, which might suggest that BAA's enthusiasm for expanding Stansted was really motivated by a desire to pre-empt capacity increases by competitors, a case of an entry deterrence strategy on the part of BAA. The position taken by the government in the most recent White Paper appears to confirm that such a strategy is still working.[15]

In the general debate on the (de)merits of the RPI-X approach, and especially the claims that it leads to underinvestment, the issue of an entry deterrence strategy on incentives to invest appears to have been neglected. But if the Stansted case is not exceptional, it suggests that the regulatory approach to sanction investment aimed at capturing market share can be a potent instrument for the dominant firm to use.

Conclusions

The expansion of Stansted in the late 1980s and early 1990s (at least on such a large scale) was most likely a mistake and one that might well have been foreseen.

---

14   Nor is it the practice to consult Heathrow airlines on the investment programme at Stansted (or vice versa).

15   In the 2003 White Paper, the government declined to support Luton's proposal for a second runway largely because its analysis indicated relatively limited traffic growth so that, given existing capacity, a second runway was 'unlikely to come to fruition for many years'. Perhaps this is not a surprising conclusion, given both the government's predisposition to expand Stansted and BAA's pricing policy at Stansted. Nevertheless, it might be questioned whether this is an even-handed approach on the part of the government and whether the decision should not be left to market forces (subject, of course, to the usual local planning scrutiny).

The project emerged from a planning process; it was not market driven. It has since been partially rescued by the low-cost airline revolution (which, arguably, it helped to facilitate). Stansted dumped excess capacity onto the market cheaply (reflecting short-run marginal costs after capital had been sunk) and the low-cost airlines responded, but it will be nearly 20 years later that Stansted's annual return covers its cost of capital (now anticipated for about 2008). During the intervening period, BAA has been allowed by its regulator (encouraged by the Competition Commission) to leverage its market power at Heathrow in order to produce a satisfactory return overall, thus compensating it for Stansted's poor financial performance following its earlier overexpansion.

There are signs that history might soon repeat itself. The government has decided that it will now support the development of a second runway at Stansted as the first major new development in the south-east. It has also stated that it will not promote or pay for this development in spite of analysis having strongly indicated that such an expansion in the time frame envisaged would not be commercially viable on a stand-alone basis. BAA argues, therefore, that it should be allowed to continue to leverage its market power elsewhere so that it can cross-subsidise further expansion at Stansted. This confronts the new policy stance of the regulator adopted as part of the 2003 price-cap decision that, in future, investment in each airport is to be judged, for the purposes of determining the price caps, on a stand-alone basis. The regulator did, however, leave the door ajar for reintroducing a system-wide RAB, indicating that it was prepared to revert to the *status quo ante* if compelling evidence was forthcoming demonstrating net user benefits and no undue distortion of the aviation market.

I have therefore examined two public-interest arguments put forward by BAA as grounds for market intervention. Both of these focus on potential user benefits not incorporated in Stansted's revenue stream. I find the first argument, concerning the relief of congestion at airports other than Heathrow, generally weak and probably unsustainable. The second argument, that cross-subsidising Stansted might benefit air passengers more generally because of the additional airline competition engendered, is a line of argument that might have had credibility in the past but not now that the low-cost airlines are well established. It also has other significant weaknesses. The chief effect at Heathrow is likely to be the transfer of scarcity rent from airlines to consumers without any noticeable impact on output: the effects would be largely distributional in nature rather than representing an improvement in economic efficiency. But, in so far as output is affected, the efficiency consequences could be negative instead of positive, with a negative impact occurring if network airlines operating out of Heathrow cut back on marginal services (both at the London airports and at regional airports), the fixed costs of which are currently sustained by profitable operations at Heathrow. In addition, there is likely to be further serious distortion of competition between airports, particularly between Luton and Stansted, the former having suffered 'material harm' from Stansted's previous overexpansion.

Reflecting upon the recent White Paper and its proposals for a further major expansion of Stansted (with an expectation of completion around 2011–12), it is difficult to appreciate the position taken by some of the stakeholders. First, it is not at all clear why BAA's management should want to continue to cross-subsidise investment at Stansted from its other airports. In the past, management has been able to pursue such an approach (and gain market share as part of an entry deterrence strategy) without prejudice to shareholder interests because of the regulator's endorsement of the system RAB approach. It is less evident that shareholders will be well served in the future by the public-interest arguments that the BAA management has adopted, even if the regulator can be persuaded to accede once again to a single RAB for all three London airports. This is because of increased regulatory risk: the regulator is not able to precommit to future policy, and compelling evidence that might lead to the re-adoption of a system approach might also turn out to be less than compelling at a later date. Whether the BAA management will maintain its current stance once existing protections from the full disciplines of the capital market have been removed remains to be seen.

It is also difficult to appreciate why the government should want to encourage an approach based on cross-subsidy, given the emphasis that is now placed on environmental considerations. The effect of the system RAB approach, and the expansion of Stansted at a rate faster than that dictated by the market, has been to encourage its use by marginal aviation activity. It is odd, therefore, that the White Paper should set out the government's environmental concerns and its determination to pursue economic instruments for mitigating environmental impacts, whilst at the same time appearing to encourage at Stansted marginal aviation activity that fails to meet infrastructure costs directly attributable to it.

Finally, it is to be noted that, in the past, the combined RAB approach to the setting of regulated prices at BAA's London airports has inhibited opportunities for reducing the scope of regulation. Stansted has never had any real market power so its price cap has not been binding. But, in order to facilitate its accelerated development ahead of market forces, it has had to be 'designated' and thus subject to the rigours of a regulatory regime. Now that the regulator has set aside the combined RAB, this provides an opportunity to reduce the regulatory burden by de-designating Stansted and allowing greater scope for competitive forces to shape its future.

Chapter 10

Airport Investment: The Regulatory Dilemma[1]

The Airports Act 1986 which privatised the British Airports Authority also imposed a condition on certain charges that BAA plc could impose on users of its three southeast airports, London Heathrow, Gatwick and Stansted. This condition takes the form of a price cap; charges are permitted to rise in line with inflation plus or minus a specified percentage, known as the 'X' factor. The RPI-X form of price regulation has been applied in a number of UK utility industries transferred to private ownership – telecom, gas, water and electricity in addition to airports. In all cases the initial condition on prices was set by the government with the requirement that it be reviewed after a specified period – five years in the case of airports. The first quinquennial review of BAA plc is now taking place: in July 1991 the Monopolies and Mergers Commission (MMC) published its findings under a reference made in December 1990 by the industry regulator, the Civil Aviation Authority (CAA).

The case is interesting for a number of reasons. First, the British style of regulating private sector utilities differs from that commonly practised in the United States and there has been a debate as to the relative efficiency of the two approaches (see Beesley and Littlechild 1989, for a good review). The MMC's review tests the arguments. Second, the BAA case is the first regulatory review of a transport industry; and, third, a crucial element in the debate has been the setting of a price cap consistent with providing BAA with an incentive to invest in new airport capacity. Although the MMC Report covered many regulatory issues this policy note focuses upon the size of 'X' in the price-cap formula and its implications for investment in a new airport terminal, Terminal 5 (T5) at Heathrow, planned by the BAA to be the next major project to increase capacity at its southeast airports.

There are a number of generic issues in the setting of X. The value of X bears upon the incentives to improve productive efficiency, achieve efficient allocation of capacity, and increase investment. The size of X is a relevant investment issue to the extent that any particular level of airport charges may be insufficient to earn an adequate return on investment. BAA currently adopts a 'hurdle rate' of 15 per cent for investment in 'core' activities (this is consistent with an estimated cost of capital of 14 per cent post tax and nominal, an assumed inflation of 5/5.5

---

1   First published in *Journal of Transport Economics and Policy*, 1991 and republished here by kind permission of the journal.

per cent, and an allowance for risk). On a project such as T5, a 15 per cent post tax nominal rate is approximately equivalent to a pre-tax real rate of 11 per cent, somewhat higher than the rate of return used by many UK nationalised industries (Civil Aviation Authority, 1991, para 13.60).

However, the return on new investment has to be set within the regulatory framework which judges X in relation to the return on average capital employed and to what is considered a reasonable return. This in turn depends upon the degree of risk faced by BAA in relation to its southeast airports. The MMC took into account BAA's monopoly in the southeast, its ownership of three airports (which allows risk to be spread) and the offsetting 'regulatory risk' associated with a succession of reviews which could change the form of regulation besides the value of X. On balance the Commission regarded the level of risk in the next quinquennium as below average. Consequently, its view, based on returns in the economy generally and the particular circumstances of BAA, is that a figure of about 8 per cent would be a reasonable return on capital employed. (It was noted that this is also the rate set by the government for investment by nationalised industries, a rate which is based on average returns on private sector assets in activities with low year-by-year variability.)

Having decided upon a reasonable financial performance for BAA, the next step was to relate rates of return to values of X. The MMC calculations for the forthcoming and subsequent quinquennials are shown in Table 10.1. The table suggests that the return on average capital employed is relatively insensitive to the value of X; this is partly explained by the large share of revenues accounted for by commercial revenues (duty- and tax-free airport sales, airport retailing generally and rental income) which do not fall within the price cap. Over the 1992/93 to 1996/97 quinquennium as a whole, the current cost accounting return on net assets would average 8 per cent for a value of X of minus 4 (RPI-4), although in two of the five years such a constraint would produce a return of below 8 per cent. In addition, the average return would fall below target for the subsequent quinquennium if RPI-4 was maintained.

Table 10.1 Southeast airports: financial results for different values of X

	RPI-0	RPI-0 (*Heathrow*) RPI-2 (*Gatwick/ Stansted*)	RPI-1	RPI-2	RPI-3	RPI-4
CCA return on average capital employed (5 year average) 1992/93 to 1996/97	9.1	8.9	8.8	8.5	8.3	8.0
1997/98 to 2001/02 (RPI-X to 1998/97 RPI-0 from 1996/97)	8.6	8.3	8.2	7.8	7.4	7.0

Source: Adapted from Civil Aviation Authority (1991), Table 13.2.

However, the return on new investment has to be set within the regulatory framework which judges X in relation to the return on average capital employed and to what is considered a reasonable return. This in turn depends upon the degree of risk faced by BAA in relation to its southeast airports. The MMC took into account BAA's monopoly in the southeast, its ownership of three airports (which allows risk to be spread) and the offsetting 'regulatory risk' associated with a succession of reviews which could change the form of regulation besides the value of X. On balance the Commission regarded the level of risk in the next quinquennium as below average. Consequently, its view, based on returns in the economy generally and the particular circumstances of BAA, is that a figure of about 8 per cent would be a reasonable return on capital employed. (It was noted that this is also the rate set by the government for investment by nationalised industries, a rate which is based on average returns on private sector assets in activities with low year-by-year variability.)

Having decided upon a reasonable financial performance for BAA, the next step was to relate rates of return to values of X. The MMC calculations for the forthcoming and subsequent quinquennials are shown in Table 10.1. The table suggests that the return on average capital employed is relatively insensitive to the value of X; this is partly explained by the large share of revenues accounted for by commercial revenues (duty- and tax-free airport sales, airport retailing generally and rental income) which do not fall within the price cap. Over the 1992/93 to 1996/97 quinquennium as a whole, the current cost accounting return on net assets would average 8 per cent for a value of X of minus 4 (RPI-4), although in two of the five years such a constraint would produce a return of below 8 per cent. In addition, the average return would fall below target for the subsequent quinquennium if RPI-4 was maintained [throughout the 1992/93 to 1996/97 quinquennium and then set at RPI-0 for the following five years].

The MMC recognised the need to consider the implications of such a charging condition for the return on new investment and particularly on that planned for Terminal 5 at Heathrow. Much of the debate on an appropriate value of X focused upon this issue. BAA believed that such a scheme could be made viable at the current level of airport charges (RPI-1) although the return could be below the 'hurdle' rate of 15 per cent. On the other hand, the implications of the T5 project, for charges in the quinquennium under review, are far from clear cut. T5 will not generate revenues until the next century, it will depend upon charges over a number of review periods, and the current review's successors will be free to alter whatever charge is set this time; the Commission remarked 'some element of uncertainty from the current regulatory system is unavoidable' (Civil Aviation Authority 1991, para. 13.80).

In response BAA argued that the current review was important in setting the 'tone' for subsequent reviews and, more significantly, that the charges set for Heathrow for the forthcoming quinquennium should be at the level necessary to earn the target rate of return (that is, the current level of charges per passenger) even though no income would be generated until subsequent review periods. In making this point it would *appear* that the BAA was pursuing the logic of its long established policy to charge on this basis of long-run marginal cost, a policy first adopted in the early 1970s (but see Starkie and Thompson 1985, for a discussion of this issue). The MMC's reply suggests that either the Commission did not subscribe to this policy or it felt that it was incompatible with the imperatives of the RPI-X approach: 'to reflect the costs of Terminal 5 in airport charges *before* the project generates revenue would require the imposition of *unnecessarily* high charges on airlines in the period before Terminal 5 opens' (Civil Aviation Authority 1991, para. 13.83 (my italics)).

Notwithstanding this view, the Commission did see merit in aspects of BAA's argument. This led it to recommend to the CAA that, in the case of Heathrow, the charging formula should allow for a limited increase in charges of 1 per cent per annum above RPI in 1996/97 on the condition that construction of the terminal starts during that year. The effect of this would be to increase slightly the average return on capital employed for the period 1992/93 to 1996/97 and, in the Commission's view, provide an adequate signal that the costs of T5 should be taken into account in future quinquennial reviews. Thus, the value of X was set in such a way that an explicit inducement to invest in new capacity was provided (see Helm and Thompson [1992] for a discussion of this type of approach). Adoption of such a condition would result in additional charges at Heathrow of about £10m in 1995/96 and £20m in 1996/97 (at 1990/91 prices) compared with revenues generated if RPI-4 were to continue to apply in those years. Finally, the Commission added that if there was sufficient support from airlines for a larger, more expensive project to be financed out of higher charger this could be taken into account during the next quinquennial review, or by the CAA considering a mid-term adjustment to the formula.

The recommendation of the MMC, that the value of X for the quinquennium should be set at –4 but on commencement of T5 the formula should revert to RPI (with RPI+1 at Heathrow), was announced in early July, at which time the CAA issued its own judgment. This rejected the MMC's advice. The Airports Act provides for the Authority to come to a different view from that of the MMC, and the Authority has given notice of the intention to set X at –8 from April 1992, with no adjustment at the commencement of T5 proposed at this point in time.

In arriving at the figure of –8, the CAA noted the MMC's conclusion that the level of risk of the southeast airports is below average, and it considered that a reasonable return for BAA plc would be 7 per cent rather than the industry average of 8 per cent upon which the MMC based the recommended value of X. An overall reduction of 1 per cent in the rate of return for the airports equates to 4 percentage points in the value of X; hence the CAA's proposal for RPI-8 rather than RPI-4. In the CAA's view this allows for BAA plc's commitment to major capital investment at its airports and produces reasonable financial ratios for the airports as a whole.

The BAA review highlights a number of aspects involved in providing the newly privatised utilities with the correct incentives for efficient investment. In particular, it highlights the difficulties of judging industry risk. Is a return of 8 per cent on average capital employed an appropriate yardstick to adopt in this case? The comparator used appears to be private sector activities with a low cyclical (year by year) variability in demand. Aviation, on the other hand, is generally regarded as a cyclical activity, and the MMC noted the important gearing effect of growth on BAA's financial performance. There is also the added risk imparted by the regulatory process itself. The MMC referred to this aspect on several occasions. The fact that its advice has since been rejected by the CAA, which has proposed its own value for X, underlines this point. Also, there is no appeal process and the CAA is able to undertake mid-term reviews.

A period of consultation and negotiation with the BAA will last until October, when the CAA will announce its final decision. It remains to be seen whether the CAA or the MMC is correct in its judgement, whether BAA plc chooses to bring forward investments as planned, and whether the unfolding regulatory process encourages the market to add a risk premium and thus increase the cost of capital to BAA. Clearly the review has emphasised one problem of the British style of utility regulation; the difficulties of reconciling a regulatory process which is essentially short term with the need to provide incentives to invest in projects that are capital intensive, firm specific and dependent upon revenues over a long period of time.

Chapter 11

Incentives and Airport Investment[1]

Introduction

The airport regulator, the CAA, is advised on setting price-caps by the Competition Commission; during the last price-cap review both bodies were reluctant to subject to detailed scrutiny the capital expenditure programme of the four price-capped airports deemed to have substantial market power (London Heathrow, Gatwick, Stansted and Manchester).[2] This reluctance is in part understandable: the relevant Statute, the 1986 Airports Act, requires that the regulator imposes minimum restrictions consistent with its functions and duties, apart from which the CAA believes more than most in regulating with a light touch. But, is such an approach entirely appropriate, statutory constraints notwithstanding; what are the nature of the investment incentives faced by price-capped airport companies and are these likely to lead to too little or too much investment?

Economic Incentives for Under- or Over-Investment

Arguments that in general price-capped firms have an incentive to under-invest and 'sweat' assets are usually based on two considerations. The first is a variant of the so-called hold-up problem: infrastructure assets generally have long lives committed to specific purposes, so that there is a risk that regulator(s) might subsequently squeeze prices to an extent that the investment is not fully remunerated.[3] The second consideration is that the regulated firm might engage in *ex-post* opportunism by reneging on CAPEX agreed as part of a regulatory

---

1   First published in *Utilities Policy*, 2006 and republished here by kind permission of Elsevier.

2   For arguments suggesting that, in general, airports might not fully exploit market power and, in so far as they do so, the impact is limited, see Starkie 2002. However, in the case of the London market, market power is accentuated both by BAA's ownership of the three largest airports and by the exceptional nature of Heathrow.

3   See Armstrong et al. 1994, 85–90. The UK airport regulator made an interesting attempt to circumvent this problem at the time of the 2003–08 review by proposing a long-term price path commitment for Heathrow and Gatwick airports. This would have entailed a 20 years commitment on price-caps linked to current capacity and that of Terminal 5, with incentives for new investments. However, the Competition Commission did not support the proposal (see Hendriks and Andrew 2004, 114).

settlement thus inflating its return.[4] These arguments have loomed large in the theoretical literature and inclined Helm and Thompson (1991), for example, towards the view that privatised utilities will tend to under-invest.

Set against these two arguments that the privatised utilities have an incentive to under-invest, are three counter-arguments, two of which take on an added significance when regulators choose not to closely scrutinise CAPEX. First, under-investing, at the same time that prices are pressed down towards competitive levels by the regulator, would mean that supply and demand have to be balanced by mechanisms other than by price alone, often by a diminution in product quality, such as by queuing. But, this will lead to a loss of reputation, the additional burden of managing congestion, disgruntled consumers and probably conflict with the regulator, possibly leading eventually to more intrusive regulation. For management, anything but a quiet life would prevail; thus, under-investment risks imposing significant additional *internal* costs on the firm.

Second, the regulator sets a limiting price, usually every five years, with reference to an allowable rate of return on assets and providing the asset-base meets, *overall*, the firm's cost of capital, the firm is able to expand its asset-base without prejudice to its return (although the size and phasing of CAPEX will have to have due regard to various financial ratios). Absent scrutiny of the capital expenditure programme by the regulator, the regulated firm is provided with an opportunity to consume its monopoly rent by expanding its asset-base and by gold plating its investments.[5] Bear in mind also that managers, when given a choice, usually prefer running large rather than small businesses; size brings status and material rewards. Chief Executives typically by nature are ambitious and large, extensive and expensive assets help to satisfy such ambitions. Such tendencies are frequently channelled into mergers and acquisitions, but in the ex-public sector utilities, opportunities, at least in the UK, of this nature are restricted by regulatory concerns.

Third, it is recognised that in imperfectly competitive markets, firms not infrequently have used as an entry deterring strategy the building-in of excess capacity (for example, Dixit 1980). Although the utilities sector can be characterised by areas of considerable market power, nevertheless, there is often a competitive fringe that can threaten from time to time the core activities of the monopolist. But (and it is a point generally ignored in the economic regulation

---

4   If the regulated company chose to game the system in this way it would seem more likely that it would seek approval for what was an exaggerated CAPEX; it is not evident that the outcome of the 'game' would be an inefficiently small investment programme.

5   Such an opportunity is exemplified by the Competition Commission's remarks at the last airport price review: 'we have not adjusted BAA's forecasts for capital expenditure: even if there is scope for lower costs on some projects, there is in our view likely to be a demand for any cost savings to be spent on additional projects' (Competition Commission 2002, para. 1:14). This has led to suggestions that in these circumstances CPI+/-X is really rate-based regulation but with a formal regulatory lag but this is perhaps too harsh a judgment ignoring for example the important forward looking nature of incentive regulation (see, for example, Beesley and Littlechild 1989).

literature), the regulated utility is well-placed also to pre-empt such entry by leveraging its market power to expand capacity through an overly generous CAPEX programme, especially when proposed CAPEX is an area treated circumspectly by the regulator. The use by the utilities of such entry deterring strategies has been noted, for example, in the European gas industry (Cornwall 2004).

Empirical Evidence on Airport Investment

What is the evidence on this general issue in relation to airports? The under-investment/asset sweating proposition would appear to receive strong confirmation from those parts of the London airports system that have been highly congested for a considerable period of time. London Heathrow is the pre-eminent example, but London Gatwick is also reasonably congested; declared runway capacity is constrained at both airports (see Box 11.1). But, it is debatable whether these capacity constraints represent a deliberate policy by BAA to limit capacity.

Box 11.1   Runway capacity and pricing

Declared runway capacity, in practice, reflects not only runway constraints but capacity limitations that exist in all parts of the airport system (see Turvey 2000).

Demand in excess of declared capacity is restrained largely by a, generally applicable, runway slot allocation process that follows administrative criteria agreed to by the European Commission (Regulation 95/93). The chief feature of this process is that, in each season's allocation of slots, prior users (in the last equivalent season) are given precedence. There is, however, a trading market wherein airlines buy and sell slots from and to each other, but, because the EC currently opposes the idea of trading, this market is opaque (a grey market).

Because airport charges are not used to balance demand and supply, the economic rents associated with capacity constraints are captured largely by the incumbent airlines and not by the BAA. If the Company was to capture the scarcity rents (e.g. by raising landing charges), these would have to be offset in some way in order to normalise the return on capital. In spite of the regulator pegging the return, BAA should still have an incentive to invest in additional capacity because it is allowed its cost of capital.

The root cause of the limitation on new investment at Heathrow and Gatwick is first and foremost the environmental/planning constraints that make physical expansion at either location exceedingly difficult. The new, very large Terminal 5 at Heathrow, which will do much to alleviate existing terminal constraints, was subject to a planning inquiry record-breaking for its longevity; proposals for extra runway capacity meanwhile are constrained at Heathrow by the difficulty of meeting European air quality standards and at Gatwick by a legal agreement, which lasts until 2019. Within this constrained framework, BAA's investment programme at the two airports has been, and remains, substantial and at both airports it is currently well ahead of projections that formed the basis of the 2003–08 price cap.

If there remains an element of doubt as to whether BAA has tried as hard as it might have done in the past to expand capacity, it is with respect to adding further runways at Heathrow. Helm and Thompson (1991) in support of their argument that privatised utilities have an incentive to under-invest, remarked that BAA saw further runway investment in the southeast as less urgent matter than the regulatory authorities, but they provided no specific evidence on the matter. More recently, the House of Commons Transport Committee (2003) took BAA to task for having told the Terminal 5 inquiry that an extra runway at Heathrow would be unacceptable for environmental reasons; they considered that this was a wilfully misleading statement.

In contrast, London Stansted is viewed as having expanded too rapidly since the late 1980's resulting in a poor commercial return (see Box 11.2). The current proposal for further early expansion of the airport also lacks a strong commercial case, in spite of which the BAA is zealously pursuing the matter. The adoption by BAA of this somewhat curious commercial policy in relation to Stansted might be explained by two inter-related factors: first, a desire to grow the company, circumventing the more severe planning constraints at Heathrow and Gatwick and second a small but nevertheless significant threat to its market share from other regional airports, in particular from plans to expand Luton airport (which, in the view of the CAA, suffered material harm from the earlier over-expansion of Stansted).[6]

Manchester Airport is also subject to a price-cap regime in spite of its status as an airport wholly within the public sector. It has had a substantial capital works programme, of nearly £400 mn in the current five year period, following expenditure of nearly £300 mn in the previous period (at 2002/3 prices).[7] It added a major new terminal in the 1990s (which is now being extended) and opened a second runway in 2001. But, whether the size or timing of the expenditure undertaken was appropriate is open to question. Although facilities were getting increasingly crowded in the late 1980s, the airport appeared reluctant to test the investment case by adopting peak load pricing, in spite of being encouraged to do so at the time of the first regulatory reviews (see Starkie 2003 for details).[8] Indeed, its ambitious capital expenditure programme of the early 1990s was such

---

6   The proposed expansion probably provides an example of an entry deterring strategy, in this case deterring expansion at nearby Luton Airport (see Starkie 2004).

7   In the review period 1998–99 to 2002–03, there was a shortfall of capital investment against the forecast. However, this shortfall occurred after 9/11 and was prompted by the dramatic reduction in aviation activity. Prior to 9/11 capital expenditure had been running ahead of forecast.

8   The airport has now, post second runway, adopted peak-load pricing.

Box 11.2   Capricious CAPEX and the system approach

The BAA forecast for CAPEX at the time of the 1997/98 – 2001/02 price-cap review included £135mn for Stansted (at 2001/02 prices), a figure dwarfed by spending planned for Heathrow in view of proposals at the latter to start construction of Terminal 5. However, BAA ran into difficulties at the Terminal 5 public inquiry, which slowed the project's progress.

BAA then simply switched a substantial amount of CAPEX intended for Terminal 5 across to the two other airports, with the result that at Stansted during the five year period, investment was 138 per cent above that originally forecast and at Gatwick 24 per cent. This raises the question of why was this additional expenditure not in the original CAPEX programme for the two airports if it was commercially viable? (Given the then relatively low gearing of BAA, any investment that met its cost of capital should not have been squeezed out of the programme by financial constraints).

This switch of CAPEX between airports was facilitated by the 'system' approach to BAA's London airports, whereby the regulated asset-base for all three airports was combined for the purposes of judging an allowable return. This had the effect of enabling BAA to leverage its market power at Heathrow in order to support under-performing assets at Stansted (for further details, see Starkie 2004). Since 2002 the CAA has changed its policy so that each airport is now considered on a stand-alone basis, but BAA is pressing for the adoption of the *status quo ante*.

---

that the then Director of Economic Regulation at the CAA commented that the airport 'was proposing a remarkable investment programme [and] gross capital formation in the last year of the quinquennium was 56 [per cent] of the forecast turnover, which is a truly astonishing figure'.[9]

Finally, in Ireland where price-cap regulation of airports was introduced in 2001, the regulator has been at loggerheads with the state enterprise running Ireland's three main airports over what he considers an excessive CAPEX programme. This strong difference of opinion was manifest at the time of setting the first price cap, with the regulator's reluctance to accept all proposed CAPEX leading to an appeal to the High Court by the airport company which was unsuccessful. Similar issues have prevailed with respect to the most recent price review. A particular concern of the regulator was the fact that the airport company proposed a large CAPEX programme without adequate appraisal or justification.

---

9   It is possible that this also reflected a strategy aimed at pre-empting potentially competing investment, in this instance by Liverpool airport.

Conclusions

On balance, taking into account both incentives and empirical evidence, it would appear that the regulated airport companies have probably inclined towards over-investment, rather than under-investment and that a price-cap regime has done little to curb the enthusiasm of the public sector in this regard.[10] This, in turn, begs the question as to whether the benign approach of the UK regulators has been entirely appropriate. It is, nevertheless, an understandable position because more detailed scrutiny requires that the regulators set their judgment against that of the airport operator in circumstances where there is an imbalance of information and expertise.

The CAA's new approach for the current 2008–13 review proposes that it should act more as a facilitator to enable the regulated companies to engage with airlines and other users in a structured way. Specifically, the CAA has proposed that the business plans for each of the four designated airports should be drawn up following detailed consultations with the airlines and that this approach should precede submission of the plans to the CAA. It is intended that the parties will agree on various key variables including, importantly, capital expenditure. If successful, the approach will move airport regulation in the direction of similar approaches used in a number of other competitive industries (not subject to economic regulation) characterised by large sunk assets where large strategic investments are secured by agreement between a firm and its major suppliers or its downstream customers, the seaport industry for example.

---

10   Alan Walters (1978) writing nearly 30 years ago in the context of a then mostly public sector industry, passed comment that virtually all large airport projects, and probably many or most small projects, are mooted many years before they are economically desirable, adding: 'airport authorities and government officials would seem to be congenitally committed to laying down new runways too soon and in too large number.'

Chapter 12

A Critique of the Single-till[1]

The Competition Commission has taken the unusual step of issuing a statement on its current thinking during the course of its review of London and Manchester Airports charges … The statement points out that the Commission has not reached a final view on the issue [the CAA's proposals for a dual-till approach] and that it remains open to further representations on the point. […] My comments follow after the quotations taken from the Commission's statement. […]

> We remain unpersuaded that in practice there would be no effect on air fares at either congested or uncongested airports if airport charges were to be higher at the three BAA London airports as a result of a switch to a dual-till regime.

Typically, airports combine provision of runway/terminal services with a retailing and property business. At large airports the latter account for a substantial turnover, which can exceed the turnover from aeronautical services. The two activities are complementary and consequently increases in air traffic volumes will usually produce significant increases in the profitability of the retailing/property activities. This provides an incentive for the airport to increase traffic volumes and has the effect of attenuating the normal downward pressure on profits that arises when prices for the use of runways, etc. are lowered to attract more traffic: airports with substantial commercial activities have good reason to limit the extent to which they exploit their market power as runway businesses. (Some airports are reported to be offering runway terminal services at negligible or zero charges in order to enhance commercial profits).

This practice is only worthwhile, however, so long as capacity exists to accommodate further air traffic. If the runways or terminals are operating at capacity then there would be little point in the airport business reducing aeronautical charges in the interests of the retailing business. At congested airports we might expect charges to rise to ration capacity notwithstanding the commercial sales.

Consequently, airport businesses have an incentive to move the basic level of charges in a direction that is generally efficient. At uncongested airports there will be an incentive to keep charges low; at congested airports the incentive will be to increase charges to ration scarce capacity.

---

1   First published in the Competition Commission's, *BAA plc: A Report on the Economic Regulation of the London Airports Companies* (2002) and republished here by kind permission of the Competition Commission.

The Commission's statement quoted above which refers to both congested and uncongested airports and to charges being higher at both following adoption of a dual-till, appears to suggest that the Commission does not recognise these differing incentives. At uncongested airports, charges (and therefore air fares) are most unlikely to be higher as a result of a switch to a dual-till regime. In fact, in the past at Stansted it has been alleged that charges have been too low. The evidence in the case brought by Luton under Section 41 of the Airports Act suggested that Stansted had every incentive to reduce charges irrespective of the single till (and that low charges were causing Luton material harm, a view with which the CAA concurred).

> There is no evidence that the single till has led to under-investment in aeronautical assets at the three BAA London Airports in the past, nor any expectation that it will do so over the next five years.

> It is not clear that the dual, as opposed to the single, till would be likely to lead to better aeronautical investment in the future.

> The dual-till could also risk unduly benefiting commercial activities, at the expense of aeronautical, which may not attract sufficient funds.

Airport commercial activities to perform well require an increasing flow of passenger traffic. Therefore we would expect a dual-till to encourage investment in aeronautical activities and, even if the dual-till did not lead to 'better' aeronautical investment in the future and, by implication, no worse investment, there would still be a gain in welfare if the dual-till lead to increased investment in commercial activities. This is especially the case if one accepts the Commission's view, expressed in para. 3 of their statement, that profits from commercial activities are monopoly rents; if so, an increase in output through further investment is to be encouraged.

I am also puzzled as to why, under a dual-till regime, aeronautical activities '... may [sic] not attract sufficient funds ...'. This can only be as a result of regulatory failure; the return allowed on an aeronautical RAB (in a dual-till context), is determined by the Regulator. Quite apart from this, the parties could enter into an explicit regulatory contract that could secure increased investment in aeronautical activities as part of an agreement to adopt a dual-till.

Similarly, the argument that there is no evidence that the single till has led to under-investment in aeronautical assets appears to ignore the possibility that the single-till might have led to under-investment *per se* by reducing the level of investment in commercial activities. Again, it is worth stressing that an increase in investment in commercial activities, *certeris paribus*, is a welfare gain. This apart, the basic premise on which the Commission has based its conclusion on this point is arguable. The single till complicates the process of determining an appropriate cost of capital for the regulated airport; combining aeronautical and retailing activities means that a view has to be taken on a composite cost of capital covering both activities. This added complication increases the risk that an inappropriate cost of capital might be chosen and that this might prejudice aeronautical investment, or commercial investment, or both. I find it difficult to agree with the proposition that the single till has a benign effect on investment in aeronautical activities.

> A move from the single till to the dual term would in the longer term mean substantial transfer of income to airports from airlines and/or their passengers ...

The degree to which income is transferred depends upon the economic efficiency consequences of a change in the regime including its impact on investment. I have already expressed a view on the investment incentives of a change to a dual-till approach noting that where there was a substantial effect on increasing aeronautical *and/or* commercial investment, airlines and their passengers would gain.

However, the Competition Commission in the above statement appears to be referring to a pure rent transfer from passengers and airlines to airport businesses. In this context, it is to be noted that a substantial proportion of aircraft movements at Heathrow are by aircraft on a foreign register, (and that a substantial proportion of passengers are not UK nationals). In efficiently operating capital markets, the distribution of rents would not really matter because ownership of the rents would be transferable. However, capital markets do not work efficiently in the aviation industry because of various institutional constraints. Shareholdings in BAA, for example, are limited to a single ownership of 15 per sent, but without any nationality restrictions; the latter restrictions do apply however in relation to airlines. Given the current use of Heathrow, for example, the effect of transferring rents from airport operator to airlines would be to transfer these rents from UK to overseas interests.[2]

I also find it difficult to agree with the basic premise expressed in the quote from the Commission. Because Heathrow and, to a lesser extent Gatwick, represent bottleneck facilities, an increase in demand for their use over time will have the effect of increasing the scarcity rents. Consequently, absent a dual-till, the size of the scarcity rents captured by the airlines will continue to increase. It would therefore be possible to introduce a dual-till without necessarily reducing the *absolute* amount of scarcity rent currently captured by the airlines. In which case, there would be no transfer in the strict sense. However maintaining existing policies will see a continuing and increasing transfer of rents to users from suppliers which appears to suggest that the Commission is giving a preference to the former party (and, given the current structure of ownership of airlines and airports, is taking a benign view of the off-shore transfer of rents).

---

2   Overseas equity holdings in BAA are about 25 per cent; for BA/Virgin it is nearly 50 per cent and for BMI 40 per cent. UK registered airlines account for about 55 per cent of movements at Heathrow. UK holdings in foreign registered airlines are minimal. Therefore, the effect of the transfer of £1 of rent from the airport operator to the airlines would be to transfer close on 50p to overseas interests.

We do not see significant benefits from any de-regulation of commercial activities.

See comments [...] above.

> We are not convinced that the current profits of those [commercial] activities should be characterised as locational rather than monopoly rents.

I find this statement surprising. I would argue that in the absence of deliberate attempts by the BAA to limit the quantity of commercial activities (as distinct from the limitations imposed by the lack of space or returns on capital), the profits accruing to BAA are of the nature of location rents. I am not aware that the BAA deliberately limits the quantum of its commercial activities except perhaps in the interests of the quality of service having regard to its aeronautical activities. The Commission offers no evidence to the contrary. If there is evidence that the BAA is acting to exploit its monopoly power in the provision of commercial services, it is to be expected that this to lead to a public interest finding. Such a finding would also signal a need to increase output and therefore investment in commercial activities and it is this that a dual-till would facilitate.

> It is not clear that the dual-till would have a significant beneficial effect on efficiency in the utilisation of aeronautical facilities, in particular, of scarce runway capacity.

> We remain un-persuaded that in practice, that there would be no effect on air fares ... if airport charges were to be higher at the three BAA London airports as a result of a switch to a dual-till regime.

These two statements are potentially contradictory. If the dual-till does have an effect on airfares at congested airports, there will be beneficial effects on efficient utilisation of scarce capacity (and less rent shifting).

If on the other hand the dual-till does not have an impact on utilisation, this is because it has no effect on airfares. In which case, the shift of (pure) rents is from airlines (with majority of foreign interest) to BAA (with a majority UK interest). However, it seems likely that there will be a beneficial effect on utilisation from an increase in charges at the congested airports. An increase in charges would induce substitution between the sub-markets of total demand (there will be a greater reduction in demand where price sensitivities are higher). Consequently, there would be more capacity available for high value users at the expense of low value users so that even if increased charges do not clear the market, they will at least have been some movement in the desired direction.

This takes on added significance in the light of investment in capacity in Terminal 5. At the moment there are less than 140 passengers per aircraft movement at Heathrow with the total number of movements around 450,000 per annum. The capacity of existing terminal facilities is about 65 million passengers per annum. A condition attached to planning consent for Terminal 5 with a capacity of up to 90 mppa, is a movement limit of 480,000. Consequently, if the capacity of Terminal 5 is to be used effectively, there has to be a substantial increase in the number of passengers per movement. An increase in charges as a consequence of a move to a dual-till regime at a congested airport like Heathrow would facilitate this objective. In the absence of a dual-till it will be more difficult for BAA to make effective use of its additional terminal capacity. Consequently, the Commission's proclivity to maintain a single till approach risks a less effective use of capacity and arguably reduces incentives for future investment.

> To the extent that some of the judgements that have to be made are arbitrary, future disputes about cost allocation could harm relations between the airport and its users.

I agree that the allocation of common costs will be, to use the Commission's phrase, arbitrary, but the effect of using the single till approach is, as we have noted above, to transfer this arbitrariness to the cost of capital.

> No useful parallels can be drawn at this time from overseas airports which use the dual-till …as their circumstances are different from those of the three BAA London airports.

The dual-till has been used extensively in Australia. However, the Commission does not explain why the circumstances are different. In a recent major review of airport regulation in Australia, by the Australian Productivity Commission, that Commission firmly supported a dual-till approach in the event of a continuation of price cap regulation. In the event, the Australian Government chose to substitute price monitoring for price capping at all Australian airports including congested Sydney.

> Nor are we convinced that the dual-till approach would act as an effective incentive on BAA to maintain or improve performance by providing 'something to lose' (through revision to a single till approach at future regulatory reviews should it fail to do so).

I presume that the Commission has formed this judgement after studying the situation at BAA Scottish airports where in effect an informal price cap applies. However, the Commission does not elaborate on this point.

> … each airport should be looked at on is merits.

I agree. Looking at each airport on its merits might lead to a different outcome for the congested airports viz. uncongested airports. For example, the single till approach might be adopted only at uncongested airports where it would be consistent with incentives for efficient outcomes. Although we would not expect such an approach to act as a binding constraint (the airport business will have an incentive to keep aeronautical charges low as explained earlier), it would, nevertheless, provide the airlines with a degree of comfort. It is at the congested airports that the single till approach has its most pernicious effects and it is at such airports that a move to a dual-till approach would be most beneficial.

In conclusion, I would make two further observations: the Commission does not directly address the point that the inevitable outcome of maintaining a single till approach at a highly congested airport like Heathrow, is that it has the effect of placing downward pressure on charges (probably leading to their reduction in real terms) when common sense dictates that charges should be rising. It is perhaps not surprising that those US airlines not currently servicing Heathrow have lobbied to be allowed entry. In so doing, they are, of course, seeking to transfer to themselves some of the scarcity rents. The Commission, in appearing to support a single till approach, increases those pressures, which might be argued to be having a detrimental effect on US/UK relationships.

Second, I would note that the Commission does not appear to address the general problem of capacity shortages in the southeast airport system. In view of the Commission's concern with competition matters it might wish to pose the question: is a continuation of the single-till likely to promote further competitive entry and thus provide competition to the current dominance of BAA in the southeast? The Commission might also consider the questions it has posed in regard to future aeronautical investment, in this context of competitive entry.

PART IV
Airport Competition

Overview

In the last year or so, airport competition matters have risen to the fore in the UK and there have been stirrings of interest in Continental Europe. Interest in the issue in the UK was driven by two specific factors. First, a feeling that, in relation to airports in general (and one or two airports in particular), the competitive landscape might have changed to an extent that less regulation was called for; second, the adoption of a separate regulatory asset base for BAA's three London airports following the 2003 Price Determination, made it easier to view each BAA airport as a separate entity and thus, in turn, made it easier to question whether all three airports should be controlled by the same company. The outcome was also twofold. On the one hand, the Department for Transport decided to revise its criteria for determining which airports should be designated and thus subject to price caps, (and to request advice from the CAA on whether Manchester and Stansted airports should continue to be designated) and, on the other hand, the Competition Commission launched a market reference inquiry into BAA. A crucial element in both instances was the nature and extent of market power in the airport industry.

Chapter 13 is based on a paper delivered at the Hamburg Aviation Conference in 2001 and subsequently published in 2002. The original intention was to attempt to stimulate an interest in airport competition matters, particularly in Continental Europe, in the hope that the subject might reach the policy agenda. Its basic theme is that economic regulation cannot truly replicate the competitive market and that it introduces its own economic distortions, so that an imperfectly competitive airport industry might be preferable to a regulated one on efficiency grounds.[1] The chapter examines the nature of competition in the supply of airport services and suggests that market power is a consequence of the difficulties associated with gaining access to competing sites rather than a consequence of high fixed costs leading to 'natural monopoly'. Nevertheless, there are opportunities for substitution between airports (and other modes of transport), although the ease of entry into the airport business and the degree of substitution of one airport for another will depend upon the characteristics of the air transport market and the historical and geographical circumstances. In much of Western Europe, for example, there is potentially a competitive landscape: the end of the Cold War has left a rich legacy of redundant airfields some of which have since been converted

---

1   This sentiment was echoed by the OFT during its 2006 study of UK Airports: '… in our view price regulation is a second best solution to competition problems' (6.30), but it is to be found also in Littlechild (1983). Littlechild saw regulation as essentially a means of preventing the worst excesses of monopoly and as a way of 'holding the fort' until competition arrives.

for civilian use.[2] Attention is then drawn to the transmittal effects of price competition in spatial markets that overlap and where the customer base cannot be segmented by service providers such as airports[3], and to the complementarity between an airport's commercial activities and its passenger volumes; both of these factors have the effect of attenuating market power. That apart, it is also suggested that the adverse economic costs of airports exercising market power in fact might be small. The overall conclusion is that *ex-post* regulation of conduct, provided for under normal competition law, is probably sufficient to curb monopolistic excesses in the airports industry; an argument that has found favour since the chapter was originally published, both with the Better Regulation Task Force (2001) and the House of Commons Transport Committee (2006). In Australia, a similar conclusion was arrived at by the Productivity Commission (2002) leading to the removal of formal price caps at all Australian airports and the establishment for the larger airports of an *ex-post* approach based on price monitoring[4]. Whilst in Canada, Gillen and Morrison (2006) in their Report to the National Transportation Act Review Committee also suggested that a competitive environment should be the focus of airport policy.

Chapter 14 has not been published previously. Its object is to refute and put to rest an idea prevalent in the UK until the 1990s and still prevalent in Continental Europe, that airports, unless large, require subsidy of one form or another; the basic economic characteristics of airports are such that for small airports financial losses are inevitable. And, of course, the supposed need to subsidise leaves the door open for a planned solution: perhaps a national airports policy with airports organised into hierarchies, air traffic directed to different airports depending upon the type of traffic, and new airports developed to fill the gaps.[5]

The chapter analyses the UK airport industry with reference to the Structure-Conduct-Performance paradigm of industrial organisation economics. It points to the diverse range of revenue earning activities conducted at airports; the

---

2   See also my Martin Kunz Memorial Lecture (Starkie 2005) on this point. The Republic of Ireland, however, would be an example of where geography and history have conspired to produce an aviation sector dominated by a single airport, Dublin.

3   See also my evidence to the CAA's Policy Issues Consultation Paper, December 2005 (www.caa.co.uk/default.aspx?catid=5&pagetype=90&pageid=6228). Subsequently, the CAA noted this as a key observation in the argument regarding airport competition (CAA 2006, 22.12).

4   Although I did not present formal written evidence to the Productivity Commission for its inquiry, I met informally with Commission staff and provided them with a number of my papers including that which forms the basis of Chapter 13. These sources were quoted extensively in its Report. A good review of the changes in Australian policy can be found in Forsyth (2004).

5   This viewpoint is summarised well in Sealy (1976, 30–35). It is a view that was consistent with a world in which air services were subject to detail regulatory controls specifying which routes could be flown, which airlines could provide the service, and sometimes which airports could be served (Article 68 of the Chicago Convention allows for this). Sealy also presents the alternative view of the competitive pricing model for airports set out in a seminal paper by Whitbread (1971).

importance of general and private aviation at many; and that for England and Wales there is a surprisingly dense coverage of airport and airfield facilities with those airports providing scheduled or charter services generally within (often well within) 1.5 hours driving time of the next nearest airport. There is some suggestion that constant returns to scale might feature and, with one or two exceptions, there is little evidence of excessive financial returns, probably as a consequence of a high degree of competition.

Importantly, however, the majority of small airport businesses in England and Wales are financially viable, and none of them receive public subsidy. In other words, the airport business is just like most other commercial enterprises; there are some fixed costs, the industry is by no means perfectly competitive, but it is competitive to an extent that market power need not be a serious concern (unless, that is, the airport enjoys Heathrow-like dominance). Thus, the quip that the emphasis now placed on airport-based retailing means that an airport is just a supermarket with a runway attached, is perhaps not too wide of the mark; in terms of models of (spatial) competition, supermarkets and airports have much in common.

Chapter 13

Competition and Market Power[1]

Introduction

The reason for regulating utility industries, such as airports, has been to curb market power. However, economic regulation introduces its own distortions and at the end of the day there is a trade-off to be made between imperfect competition and imperfect regulation. Because the balance between the two imperfections can change, the need for economic regulation should be re-examined from time to time. In the airport industry, circumstances have changed in recent years in a way that suggests that competition is now more important. The chapter then considers market power in the airport industry and suggests that this stems largely from the problems of gaining access to suitable competing sites and from the agglomeration economies of the air service network. Nevertheless, there are opportunities for substituting different airports but the degree to which this is possible depends upon the particular circumstances. The chapter then puts forward a model of airport competition in a spatial context that suggests that price competition can extend beyond overlapping catchment areas and that substitutability is too narrow a criterion for judging market power. There is also some competition from surface modes of transport and international air service agreements can sometimes constrain the ability of airports to exploit their market power. Attention is then drawn to some structural features of the industry that are unusual. It is suggested that these provide an incentive for airports not to exploit their market power. It has also been suggested that even if airports exploit their market power, the adverse economic costs might be small. Finally, the paper concludes by considering again the balance to be struck between the economic effects of some market power and the potential distortions of regulatory intervention.

Imperfect Competition or Imperfect Regulation

Competition is generally accepted as being good for the economy because it encourages firms to be cost-efficient; it drives down prices and leads to expanding output. It is, therefore, paradoxical that in certain circumstances production by a single firm with no direct competitors can result in even lower unit costs than production by the many. These circumstances occur when economies of scale

---

1   First published as 'Airport Regulation and Competition', *Journal of Air Transport Management* (2002) © D Starkie.

and scope result in average costs of production falling over the entire range of output that the market can absorb. When this is the case, production by the single firm is not only potentially cost efficient but it is the only sustainable industry structure. But, of course, the single firm, because it is no longer faced with direct competition, has an incentive to restrict output, raise prices and achieve a level of profit in excess of that required to provide a satisfactory return on capital. The *raison d'être* of economic regulation is to prevent this happening: to try to encourage the 'natural monopolist' to be cost efficient and to increase output to a level that maximises economic welfare. The best forms of economic regulation try to achieve this by providing the (monopoly) firm with incentives and, as Train (1991) reminded us, the central issue of regulatory economics is the design of such incentive mechanisms.

The task is, of course, easier said than done. If the Regulator had perfect knowledge about the firm, he or she could simply require it to produce a specified level of output from a specified set of inputs and sell this output at a specified price. But such complete information is not available to the Regulator who, at best, has only a partial insight into the firm's inner workings (and is also probably less informed about the characteristics of the market). Instead, the Regulator must establish ways of inducing the firm, in its pursuit of profit, to converge to the socially optimal level of output. Unfortunately, a little knowledge can be a dangerous thing; the incentive mechanisms themselves can lead to distortion and unnecessary costs. An example of this is the tendency for the price-cap form of regulation (commonly used in the UK) to encourage the degrading of product or service quality (Rovizzi and Thompson 1992).[2] In turn, this can lead to further regulatory intervention, to complex regulation (possibly with significant compliance costs) and to increased regulatory risk that has the effect of increasing the cost of capital.[3] At the end of the day, therefore, there is a trade-off between living with imperfect regulation or with imperfect markets. It is only when the market does not work well, when there is a clear case of natural monopoly *and* when regulation can reasonably be expected to improve matters that the regulatory option is worthwhile. Market imperfections alone are not a sufficient justification for intervention. Moreover, once economic regulation has been introduced in a particular sector, the case for having it at all should be re-examined from time to time in the light of changing circumstances.[4]

---

2   An unregulated monopolist might find it profitable either to oversupply or to under-supply quality, depending upon demand conditions. Once the firm is subject to a binding price cap, it will always be profitable to set quality below the efficient level.

3   As the UK telecoms regulator has noted recently: 'Setting a network charge cap is ... extremely difficult in the current climate and the risk of serious error is significant' (OFTEL Price Control Review, October 2000).

4   The legislation regulating Australian airports explicitly allows for this.

Changing Circumstances

Scheduled airline services operating on international routes are still largely controlled through a system of bilateral air service agreements established after the Chicago Convention of 1944, the better known example of which is the Bermuda Agreement between the UK and the US.[5] These bilateral agreements have the effect of stifling route innovation, they effectively ban price competition and they encourage the sharing of markets and the pooling of revenues. The outcome is an absence of competitive pressures driving down costs; high costs are largely passed through to the passenger by the colluding, often state-owned airlines.[6] Importantly, this collusive behaviour has also reduced the incentives for airports to compete. The straight jacket of the bilateral agreements means that services cannot be competed away from those gateway airports specified in the agreements. This has encouraged a complacent attitude by many airport managers. As Barrett (2000) has remarked, the world of non-competing airlines has been mirrored by a world of non-competing airports.

But the world is changing and two factors in particular have combined to bring about this change. One factor is increasing airline de-regulation that has opened up markets to both potential and actual competition between airlines and between airline alliances (Button 1991). Competition between airline alliances has focused upon connecting services through hub airports and this has led *de facto* to increasing competition between such airports. But, other airports too have become part of the competitive strategy of airlines, particularly since the advent of low-cost carrier competition. As a result of airline competition, fares, especially economy fares, have fallen and, as Barrett (2000) has also pointed out, this has tended to increase the importance of airport costs in the average fare.[7] The second factor, running in parallel to the first, and encouraged by it, has been the commercialisation of the airport industry. Airports, many of which have been treated in the past as public service organisations directly controlled by government administrations, have increasingly been restructured as public enterprises or, in a number of cases, have introduced private capital, or have been privatised. This, in turn, has led to a much more competitive outlook on the part of airport managers, who have sought to attract the now competing airlines.

---

5   A good review of the setting up of the Bermuda Agreement will be found in Mackenzie (1991).

6   One important exception to this scenario was the charter airlines that accounted for a large proportion of European air travel. This sector was competitive and cost conscious.

7   Although airline competition has had a greater impact on economy fares, particularly those carrying restrictions, on the other hand there is some evidence that the low-cost carriers are beginning to eat into the business market because business travellers appear to be becoming more cost conscious (see Mason 2000).

Sources of Market Power

This change in circumstances has, in turn, called into question the conventional wisdom that views the airport industry as an example of a natural monopoly industry not capable of sustaining competition (and thus requiring regulation). This view in large measure stems from the belief that as airport capacity is increased, long-run average costs fall.[8] However the evidence for this is at best equivocal. Kunz (1999) has pointed out that the coexistence of several terminal operators at a single airport indicates that there are no significant scale economies whilst Doganis (1992) suggests that if airports do experience economies of scale it is probable that it is only small/medium sized airports that do so. As for large airports, it seems more than likely that the average cost of expanding capacity is increasing, rather than decreasing, as one would normally expect to find in a natural monopoly industry.[9] This is because there comes a point when it is increasingly difficult and expensive to design, build and operate facilities that coordinate (spatially and functionally) activities across an expanding area.[10] Airports are perhaps a good illustration, therefore, of the law of diminishing returns, with the fixed factor in this instance being inputs of centrally located land. The argument was made some time ago (Starkie and Thompson 1985, 47–52) but econometric evidence is now beginning to support this supposition. Pels (2000), for example, found that a number of large European airports (Rome Fiumicino, Frankfurt, Munich and Zurich) were all operating under decreasing returns to scale, whilst others (Amsterdam, Brussels, Manchester, Paris Orly and Stockholm) showed partial evidence that this was also the case.[11] If there are diseconomies of scale in the airport business and the incumbent airport is large and operating at or near to capacity, there should be no barriers to prevent competitive entry. In circumstances, where there are decreasing returns to scale, the entrant can enter at a lower level of average cost than the large incumbent, compete effectively and, eventually, drive the incumbent from the market (or to operate at a reduced scale). But, if this were the case, then, *ceteris paribus*, we would expect to see no large scale, congested airports; only airports of a moderate size operating close to their

---

8   This is a view expressed, for example, in the UK DTI's March 1998 Green Paper on utility reform.

9   This is not to deny that there are significant economies of density and utilisation because of lumpy investments, although the extent of the indivisibilities in the airport industry has probably been exaggerated.

10   The complexity of the flows of passengers at a multi-terminal airport, such as Heathrow, is well illustrated in Hanlon (1996, Fig. 5.4).

11   This has implications for the regulatory approach. It suggests, for example, that prices in excess of average costs are not necessarily inappropriate, even in the absence of runway or terminal congestion. And it also suggests that basing regulated prices on normal or reasonable rates of return on capital could lead to inefficiently low prices (Starkie 2000a). Because the latter has been the usual practice in the case of the regulated British utilities, it perhaps indicates the distortions that regulation can introduce. In this instance, the effect could be to reduce incentives to expand capacity.

optimal scale of output. This is clearly not the situation and, therefore, there must be barriers other than returns to scale that prevent, or make, entry difficult.

One such barrier is gaining access to a factor-input, namely land, essential for the establishment of a new airport. As existing airports grow and become big, they attract complementary activities (airfreight distribution centres, leisure industries etc.) and these, in turn, attract a resident workforce with its supporting urban infrastructure. This pushes up the opportunity cost of land in the vicinity of the existing airport and, consequently, the costs of land assembly for new runways and terminals. It also means that there are increasing costs of noise, air pollution and congestion, which, on the whole, are not borne by the incumbent airport business. This raises the level of political opposition to new airport construction or, if the entrant airport business seeks to internalise the environmental costs to obtain acceptance by the community, it increases the costs of entry.

Although, for these reasons, the cost of entry into the airport business is probably increasing over time, innovation and developments in technology are helping to counteract this cost trend. One can note, for example, improvements in aircraft design (which are reducing the required lengths of runway and reducing aircraft noise and pollution); the development of high speed ground transport networks (so that airports can effectively serve the market whilst located further from its geographical centre);[12] and the innovation of spatially separating terminals and runways, placing the former closer to the market than the latter.[13] In addition, reductions in military spending, especially since the ending of the Cold War, have led to written-down, airfield assets becoming available.[14] Bearing in mind that many of the airports in the UK, for example, are former military airfields (including Heathrow and Stansted) this development is significant. An important barrier remains, however, if it is necessary for the entrant to assemble land to develop a completely new airport.

Another source of market power in the airport business is the agglomeration economies associated with network externalities. Both airlines and passengers gain from a concentration of air services that feed traffic to and from each other. A flight from hub airport A to airport B can also carry passengers who have transferred from in-bound flights coming from airports C to Z. Airlines gain from concentrating services at a transfer point because it permits the use of larger and more economical aircraft, (albeit over shorter and less economical average stage-lengths). Passengers gain from increased frequency and network scope

---

12   A recent example is the development of a new airport aimed at budget carriers at Cuidad Real, 200km south of Madrid but with an anticipated journey time to the city of 50 min by the new AVE high speed train.

13   Arguably, Heathrow already has its Terminal 5. It is located at the baggage check-in area at Paddington Station, the terminal for the Heathrow Express train service to the airport.

14   For example, a UK property company, Peel Holdings, has purchased a former RAF base, Finningley near Doncaster. Manchester Airport, which recently acquired nearby Humberside airport, has indicated its intention to oppose the development of Finningley as a civil airport, presumably because it poses a threat to Humberside.

and, thus, from a greater range of choices (Tretheway and Oum 1992), although this is offset to some extent by more indirect routings.[15] The significance of these agglomeration economies/network externalities is that they tie the individual airline to the hub airport and make it more difficult for rival airports to attract airlines and passengers through price competition. However, as we have noted, airline alliances do compete with each other over their respective hubs with the consequence that there is a degree of competition (albeit indirect) between the hub airports.

Finally, we can note that in addition to the entrant airport having to sink costs on entry, this is true also for those airlines (which are further down the supply chain) that move their operating base (or split their operations over more than one airport location). This factor also gives the incumbent airport an added advantage and thus increases its market power.

Opportunities for Airport Substitution

Connecting Services

If the market power of an airport is explained in part by agglomeration economies associated with its network of air services, the more connected or 'networked' these services, the more dominant the airport is likely to be. It is likely to be most dominant when it is acting as a major hub and when many passengers are transferring between flights (either as on-line passengers or as interline passengers).[16] Consequently, for a scheduled carrier, with a high level of transfer passengers to and from other airlines (British Midland at Heathrow for example), to choose to forego the revenue and cost advantages of the hub by substituting a proximate, even adjacent, alternative airport would seem most unlikely. However, even in this context of highly networked air services, an airport's ability to set prices is constrained.

This is because hub airports and their airlines compete with each other for long-haul transfer traffic. In Europe, for example, London, Amsterdam, Paris and Frankfurt are the chief competing hubs and transfer traffic can account for a sizeable proportion of their total traffics. Transfer traffic is considered to be sensitive to different price/frequency combinations offered via different hubs, although no empirical data illustrating this is to hand. If this is the case, it would suggest that in this part of the market, hub airports cannot set prices without regard to the consequences for their resident airlines which are competing for transfer passengers against airlines located at other hub airports.

---

15   Studies in the US market have shown that on balance passengers have gained, e.g. Morrison and Winston (1986).

16   In 1998 approximately one-third of traffic at Heathrow was transfer traffic of which about one-half was interlining between carriers and half was on-line transfers between aircraft operated by the same carrier (nearly all BA). See GRA and Economics-Plus (2000).

The picture is complicated, however, by the presence at major hubs of substantial originating and terminating traffic. In this originating and terminating part of the market, airlines have greater market power than they do in the transfer market. This they can usually exercise through market segmentation (charging different fares to different passengers on the same flight) but airports are more limited in their ability to adjust charges in the same way. The landing aircraft will carry passengers from a variety of market segments (including the time sensitive business traveller and the more price conscious leisure traveller) and it is not evident that the airport can levy charges that distinguish between these.[17] But, if airports were able to price discriminate between passenger markets, this would limit the restraining effect of hub competition on their exercise of market power.[18]

The market power of a hub airport might also be restricted if it is dominated by one airline and if there is the possibility that such an airline could de-camp all or part of its operations to an alternative (proximate) site. This it might be able to do without too serious an impact on its traffic volumes.[19] Its protection lies in its size and, therefore, in its critical mass. The ability to carve out traffic in this way is illustrated by British Airways project *Jupiter* in the mid-1990s; the plan to shuffle air services between Heathrow and Gatwick. Specifically, a number of BA's long-haul routes, especially those to Africa, were moved from Heathrow to Gatwick with the intention of freeing slots at Heathrow for other services.

A change of strategy following a decision to place less emphasis on low-yield transfer traffic, has since led to a further re-shuffling of routes between the two airports. Of course, in this example, the two airports concerned are in common ownership and each is very congested, so that room for manoeuvre is limited. But it is an example that does illustrate the point. If Heathrow and Gatwick were not in common ownership, and at least one of them had spare capacity, then BA (a strongly networked airline) might have a greater opportunity to substitute airports and, consequently, might have significant countervailing power.

Point-to-point Services

Opportunities for substitution between different airports increase once we take into account non-networked air services operated by the charter carriers and by low cost carriers.[20] Services by the charter carriers usually fall outside of the

---

17   Airports do discriminate in their charges between the broad categories of transit passengers (those who fly in and fly out on the same aircraft), domestic passengers, and passengers on international flights.

18   Insofar as airports were able to exploit market power in the passenger O&D market, this might have the effect of transferring scarcity rents from airlines at congested airports.

19   [This] would also apply to a group of airlines if they were able to coordinate their response.

20   Seventeen airport case studies in this part of the market will be found in Barrett (2000).

remit of the bilateral system and they have long been important in European aviation traditionally serving the inclusive tour market. Because the inclusive tour passenger is price sensitive (and not connecting between flights) the charter airlines as a consequence have had more scope for switching operations between airports in order to reduce costs. However, since deregulation in Europe, the distinction between the non-scheduled charter market and the scheduled sector has disappeared and the newly established low-cost carriers have expanded into and enlarged the price sensitive part of the market. In doing so, low-cost carriers such as Ryanair and easyJet, have rewritten the rulebook. Because they eschew the IATA interline system (enabling passengers to transfer between carriers using a single ticket) and because their average (single-class) fares are particularly low they tend to create their own market. To a large extent, they are no longer selling a product differentiated by qualities and thus having the characteristics of 'new commodities' (Ironmonger 1972). Instead, the emphasis is on price and quantities consumed. Because of this emphasis, low-cost airlines seek out airports that will minimise their operating and station costs and then create the market around them.

Airports appear to compete vigorously for the custom of low-cost airlines both as a flight destination and also as an airline-operating base. Competition between Luton and Stansted in the early 1990s for the custom of Ryanair provides an example of the latter. Encouraged by the attraction of low charges, Ryanair switched most of its flights from Luton to Stansted. This led to the former bringing a case under Section 41(3) of the UK Airports Act arguing that Stansted Airport's charges were predatory in intent. The CAA did find that Stansted's pricing policy had caused Luton material harm (but was unable to put forward a suitable remedy). Stansted had been included in the price cap for BAA's London Airports following the 1986 Airports Act (for reasons that are not entirely clear) but either the airport was discriminating in its level of charges or the price cap was not binding on Stansted.

There are also examples of normal scheduled carriers operating point-to-point services having similar opportunities to use alternative airports. Such opportunities clearly depend upon the circumstances but, if history and geography have combined to produce proximate airports, airport market power in this segment of the market could be severely curtailed. An example is provided by the Northern Ireland aviation market wherein there are two airports serving Belfast, Belfast International and Belfast City Airport. Belfast City Airport which as its name suggests is closer to the city centre than the former, opened to commercial traffic in the early 1980s. It was previously a private facility used by Shorts (the aircraft manufacturer) for testing aircraft and it expanded rapidly as a result of airline services in the short-haul, largely domestic, markets. There was strong evidence that, by the time that Belfast International agreed a take-over of Belfast City Airport, the latter had taken market share from the former. The merger was subject to an investigation by the competition authorities, which concluded that the two airports had competed 'vigorously'. Subsequently, the government blocked the merger on competition grounds. However, we should also note that runway limitations prevent Belfast City Airport from handling larger jets (without payload penalties) so that it could not compete with Belfast International in long-haul markets (a situation which Belfast International could exploit through a system of weight-based charges for the use of its runways).

Airport Competition: An Overview

The preceding review suggests that the market power of an airport with respect to its aeronautical charges is likely to vary between different, and possibly fairly narrow, segments of the air transport market. In general terms, an airport is likely to have most market power in relation to networked airline services wherein economies of scale and scope are pronounced. It is likely to have least market power in the low-cost carrier, point-to-point and inclusive tour charter market. But, importantly, in all markets the degree of market power that an airport has will be determined largely by the availability of proximate airports that are able to act as close substitutes. Consequently, the answer to the question: 'how much market power does an airport have' is circumstantial; it has to be answered on a case by case basis.[21]

Are there, nevertheless, any guiding principles and is there an appropriate framework for the examination of airport competition? It would appear that the traditional argument for economic regulation – that the airport industry is a natural monopoly industry – is no longer appropriate. It is not evident that airports are a decreasing-cost industry, at least not over more than a modest range of output. Even so, there are impediments to competition. Consequently, a more appropriate framework for analysis is to view the airport industry as an industry subject to imperfect or monopolistic competition in a spatial setting.[22]

A distinguishing property of such competition is that the seller (in this instance of air services) has some power to choose the price. This requires that consumers can be assigned to separate markets and that the individual firm can exercise some local monopoly power over one or several of these. The basis for assigning consumers to separate markets is the spatial setting in which consumption takes place. Nevertheless, a firm having a local or regional monopoly is still subject to competitive pressures. Firms share common boundaries for their markets and are thus directly affected by spatial competition. Importantly (as Greenhut et al. (1987) have pointed out), for competition to be effective firms need not compete for all of the same customers; *competition at the boundary points* is *often sufficient to transmit price changes over the whole of the market.*

---

21   The usual approach in competition analysis would be to define the relevant market for an airport's services in both its product and geographical dimensions and, thus, define the degree of competition that the airport faces. There is a body of case law developed under European competition legislation that addresses this question. In defining the market in these cases, the competition authorities have defined the relevant markets quite narrowly with the analysis frequently conducted on a route by route basis. For further details see Soames (2000, 204–11).

22   The following paragraph draws heavily upon ideas in Greenhut et al. (1987).

The point is illustrated in Figure 13.1. This shows hypothetical catchment areas for two airports (A and B), that overlap in the heavily shaded area. Within this latter area, air services from the two airports compete directly for customers. However, passengers located at point Z (well outside the catchment area of airport A and thus captive to airport B) are potential beneficiaries of the price set for customers located in the area of direct competition. Unless it is possible to separate passengers at Z into a market separate from those in the overlapping zone, the former passengers when using airport B will also benefit from the competitive price offered to passengers in the overlapping catchment areas.[23]

If there is a uniform population density over the economic landscape, we might expect airports to locate so that their catchment areas filled economic space (a Losch-type landscape) with common boundaries between their respective catchment areas ensuring competition throughout.[24] The size of the catchment areas would be determined by the depth of the market (the density of consumers) and the production technology (with different types of market leading to a hierarchy of airport types effectively with different production technologies, e.g. runway lengths). Consistent with Chamberlin's (1933) theories of monopolistic competition, super-normal profits would attract new entry, with the size of the catchment areas reduced until normal profits were established. If super-normal profits continued to exist this is more likely to be due to the problems of gaining access to suitable sites at central places rather than to economies of scale (which we have previously argued are likely to occur only up to moderate levels of output).[25] This suggests that airport competition policy should place emphasis on encouraging new airport developments and easing the problems of entry and that judgments regarding market power should extend beyond the consideration of whether airports are directly substitutable.

Modal Competition

There is also strong competition in the short-haul air transport market from surface transport modes, road and rail. Competition from the former surface mode is significant but less well-documented than from the latter. Mandel (1999) included both surface modes in his simulation of the impact of an additional DM50 charge at Hamburg Airport, which had the effect of decreasing its traffics by 8 per cent. Surface modes gained about three-quarters of the passengers no

---

23   The way that competitive effect are transmitted can be quite complex as Mandel (1999) has shown. This is particularly the case when passengers are using feeder flights from other airports.

24   Barrett (2000) quotes an interesting statistic which suggests the semblance of a Losch-type landscape of airports. In France, the UK and Germany there are, respectively, 32, 34 and 38 airports within one hour's surface access of another airport.

25   As Waterson (1984, 115) points out some degree of increasing returns are necessary otherwise firms would produce at points where consumers are located and the spatial context would become vacuous.

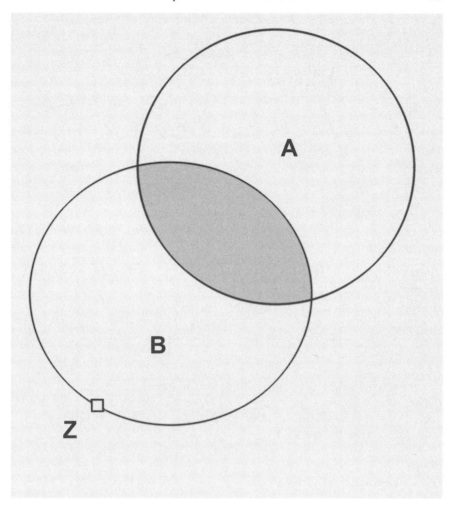

Figure 13.1 Competition and catchment areas

longer using Hamburg Airport after the price increase. Airports such as Bremen, Hanover, Kiel and Copenhagen were net beneficiaries but other airports that had relatively large traffic flows to and from Hamburg, such as Frankfurt, Dusseldorf and Munich, were also significant losers. The impacts from changing the balance of competitive advantage are therefore quite complex.[26]

A recent example, indicating fairly high cross-elasticities between air travel and rail, is provided by the UK. Following a derailment in October 2000 caused

---

26 Reflecting upon previous arguments, if Hamburg had been able to discriminate in the imposition of the hypothetical additional charge so that passengers living close to alternative airports were charged less, its overall loss of traffic would have also been less..

Table 13.1   Month on month growth of UK domestic air passengers, 2000

	Routes to/from London (%)	Regional routes (%)	Total (%)
January	6.20	5.40	5.70
February	9.10	7.20	7.90
March	5.50	4.80	5.00
April	6.90	5.10	5.80
May	7.20	8.00	7.70
June	2.30	4.00	3.40
July	4.00	4.00	4.00
August	4.50	4.60	4.60
September	1.50	1.20	1.30
October	2.70	4.20	3.60
November	14.20	11.70	12.60
December	14.30	10.30	11.80

Source: CAA

by a broken rail, severe speed restrictions were imposed on the whole of the UK rail network. The impact on domestic air travel was immediate and significant. Table 13.1 shows the monthly pattern of growth on domestic air services with a spike occurring at the end of 2000.

Other Constraints on Market Power

There are other constraints on the market power of airports that arise from the framework of international agreements that govern world aviation. An example of such a constraint is Article 10 in the UK/US Air Service Agreement (ASA) which places obligations on the respective governments to ensure that airport charges are both just and reasonable. In the late 1980s the US government brought a case before an arbitration tribunal concerning charges at Heathrow Airport arguing that the *level* of the charge for international passengers was essentially discriminatory from the point of view of US airlines. The outcome of this long and detailed case was that the tribunal ruled in favour of the US position and substantial damages (of the order of US$ 30 million) were paid to Pan Am and TWA. Not all ASAs include this type of explicit constraint but where they do, the airlines are presented with a credible threat which they can use should an airport be tempted to exploit its market power.

Why Market Power Might not be Exploited

Thus far I have suggested that although airports have a degree of market power this will vary according to circumstance. In some instances this power will be significant whilst in other instances the possibilities for substitution between airports will limit the ability of airport management to set prices, particularly in some segments of the market. But there is in fact a wider question: even if an airport has market power and dominates its (local, regional or national) market, will it choose to exploit this market power? In normal circumstances this might seem a peculiar question to pose. If the airport owners pursue profit maximising policies then we might expect market power to be exercised by aeronautical charges being set at a level that leads to abnormally high returns on capital. There might, however, be good reasons to suppose that this will not be the case.

An airport, at its most basic level, is a facility to enable passengers to join or leave aircraft and for aircraft to take off and land. In practice, airports usually combine the runway/terminal business with an extensive retailing and property business, with the largest ones combining shopping malls (for passengers), with extensive office, maintenance and cargo facilities. In the process, they have become diverse businesses but, importantly, ones that bundle together activities with complementary demands. Equally important as Kunz (1999) pointed out, some of these activities enjoy locational rents. Just as retailing property in prime locations command high (quasi) rents from superior locations, equally retailing outlets at Heathrow, or any busy airport, can achieve high returns for the landlord. And, because retailing and property activities gain these locational rents, increases in traffic volumes at an airport will often produce significant increases in their profitability. Therefore, for a profit-maximising airport company with market power, the effect of the demand complementarities linked to the locational rents, is to attenuate the normal, downward pressure on profits that arises when increased air traffic volumes are at the expense of lower prices for the use of runways and terminals. This means that, as long as an airport combines both runway and retailing activities, it will have an incentive to increase airside output beyond the point where marginal revenues from aeronautical charges cover airside marginal costs; charges will be lower than if runways were a stand-alone facility. In other words, airports have good reason to limit the extent to which they exploit market power.

The point is illustrated in Figure 13.2, which in the left diagram shows in stylised form the demand (D_1) for runway use and a corresponding marginal revenue curve (MR_1). For ease of exposition, it is assumed that there are constant costs of production, i.e. marginal and average costs are equal over the range of output considered. Given these conditions, a profit maximising stand-alone runway business will price at P_1 and produce O_1 output. In the right diagram are shown the demand conditions faced by a complementary business that owns airport retail facilities and awards franchises on the basis of turnover related fees. With airside output of O_1, the demand curve for franchises is D_1. Again, marginal costs are assumed to equal average costs. The price charged for franchises is P and retail sales are S_1. However, in the circumstances shown, it would be in the

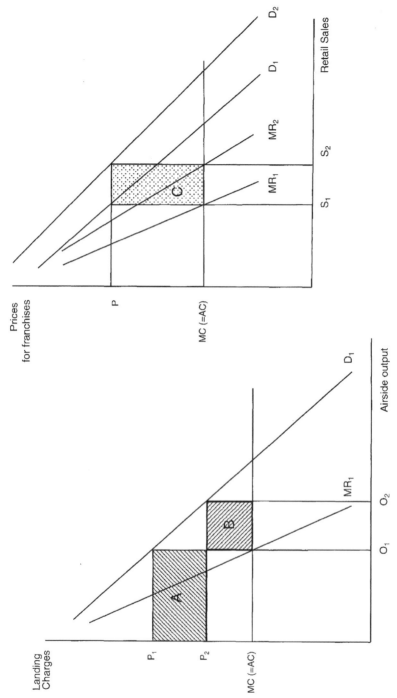

Figure 13.2 The effect of location rents on airport charges

Note: For an explanation, see text.

interests of the retailing business to make side-payments to the runways business to encourage the latter to expand output beyond the level O_1. However, the opportunities for mutual gains are exhausted once airside output is increased to O_2. At this level of airside output, the demand curve for franchises has shifted to D_2 and the location rents extracted by the retailing business will have increased to the extent of the shaded area C. The latter area is equal to (and offset by) the (negative) change in producer surplus (A less B), as a result of the runway business increasing its airside output from O_1 to O_2. Thus there are incentives to expand output until the gains to the retailing business are cancelled out by losses to the runway business. In practice, of course, airports are an integrated business with an interest in both the runway and retailing markets. The integrated business will face the same incentives to expand airside output up to an output level O_2. The important point is that the latter exceeds the level of output produced by a profit-maximising stand-alone runway business.

For an airport to have an incentive to increase airside output in the above fashion, certain conditions apply. First, airport management will need to be motivated by profit, revenue or output maximising objectives, objectives that provide an incentive to steer more customers through the retail outlets. In contrast, if these objectives are absent or compromised by, for example, limitations on share ownership, management might not have the same incentives to maximise shareholder returns by growing the retailing side of the business. In turn, they would have less interest in boosting passenger numbers by attenuating aeronautical charges. Paradoxically, it is the efficient pursuit of profits that is most likely to result in an attenuation of airside charges below monopoly levels. Second, the location rents must be of sufficient magnitude to justify the reduction in aeronautical charges in the first place. This will depend upon a number of factors, including: the elasticity of passenger demand with respect to airport charges; the size of the passenger *related* spend in airport retailing;[27] and the operating margins in the latter business. In turn, these factors will be influenced by the characteristics of the air travel market. Long-haul passengers tend to have a higher spend in the retailing outlets, particularly if they are able to take advantage of duty free sales.

If Airports are Monopolies, Does it Matter?

Earlier I have suggested that there are opportunities for airport substitution, but the degree to which substitution is a feasible proposition depends upon the particular circumstances – the type of traffic the airport handles and whether, as a result of history, geography and enterprise, there are proximate alternative airfields. Sometimes history and geography means that there are no opportunities for substitution. Forsyth (2001) suggests that this is the case in Australia where, except in minor ways, there is little likelihood of competition between airports;

---

27   Persons accompanying departing passengers and those meeting and greeting passengers also purchase at airport retail outlets.

airlines cannot credibly threaten to take their business elsewhere. However, he does go on to suggest that if an airport did charge high prices relative to its costs, the dead-weight economic loss from doing so is likely to be low. This is because the impact of higher charges on the quantity of services demanded will be small due to low price elasticities of demand. In addition, if airport charges were proportional to aircraft weight (the traditional charging approach) this would reduce the impact on demand still further. This is because heavier aircraft operate over longer distances for which passengers pay higher fares. Consequently, airport charges can be increased proportionately with aircraft weight without increasing as a proportion of the passenger fare. Overall, it is suggested that the efficiency costs of the (unregulated) airport monopoly might be small.

We should note, however, two qualifications. First, and one noted by Forsyth, is that the small efficiency cost of an unregulated monopoly assumes that the airport businesses do not dissipate rents from higher charges in unnecessary costs and excessive investment. We would assume that this would not occur if the airport company was truly seeking to maximise profits; for any given level of output profits are maximised when costs are minimised.[28] Second, the argument is based on the presumption that price elasticities are low, but the information on this point from analytical studies is ambiguous. Although estimates have been produced, it is unclear whether the resulting statistics represent the effect of a price change on the aggregate volume of air traffic, which is of relevance here.[29] However, we should also note that the higher the elasticities and thus the greater the potential dead-weight losses from monopoly pricing, the more likely it is that the airport will attenuate its charges in order to increase its retail sales and profits.

Conclusions

To conclude, we can pose the question: is *ex-ante* economic regulation of airports necessary? As was pointed out at the beginning of the paper, there is an important trade-off to be made between the economic effects of (some) market power and the potential distortions that specific industry regulation can introduce. These distortions will depend upon the type of regulation introduced; they can be minimised by well-designed regulation but they are unlikely to be avoided entirely.

Take, by way of example, the single-till approach to price-capping airport charges: forecast revenues from retailing and property activities are netted from the airport's total revenue requirement leaving only a residual revenue requirement to be covered by airport charges.[30] This introduces a series of potential economic

---

28 Perhaps this is a further reason for privatising airport utilities.

29 Oum et al. (1992) surveyed a number of studies and concluded that: '... we believe that the demand elasticity of business air travel is less than unity, while that of holiday travel is greater than unity, although the empirical estimate is not unambiguous.'

30 This single-till approach is similar to the so-called residual approach in the US.

distortions. It changes the balance of investment incentives between airside activities and commercial activities; it requires the Regulator to make difficult judgements regarding the cost of capital; and it encourages perverse pricing signals at congested airports (CAA 2000; Starkie and Yarrow 2000; Starkie 2001b). But, on the other hand, if one seeks to address these problems by trying to align charges with the costs (both current and capital) of airside activities (the dual-till approach), further problems are encountered. These include: how to treat costs which are common between airside and commercial activities, and what facilities to include as essential for the provision of air services (being those that embody the core market power of the airport).

Because *ex-ante* economic regulation is complicated in this way, it suggests that the adverse economic effects of market power need to be significant to make the regulatory approach worthwhile. For reasons explained in this paper, it is not evident that for airports this is the case. Whilst one can argue about the extent of potential substitution between airports (or other modes) in particular circumstances, perhaps in this industry there is a more important factor restraining market power. This factor is the complementary nature of the demand for air services and for retail spending by passengers, together with the existence of location rents. This, it has been argued, provides an important incentive for airport management to keep charges lower than they might otherwise be: airport businesses have a built-in incentive not to exploit their market power as airports *per se*.[31]

This is not to suggest that in this industry competition is of little consequence. On the contrary, competition matters if only to provide an additional layer of defence against the possible abuse of a dominant position. And because competition matters, there is a need for a more cogent policy on issues such as airport mergers and acquisitions, state aids to the airport industry, ownership of airport systems and the formation of airport alliances. In the meantime, whilst these issues are being addressed, I would suggest that it is appropriate to start from the presumption that the airport monopoly can be handled satisfactorily within the' context of *ex-post* regulation of conduct provided for by normal competition law; *ex-ante* economic regulation is probably unnecessary.

---

31 Note that market power *per se* is not illegal but it is the abuse of a dominant position that transgresses the anti-trust regulations.

Chapter 14

The Financial Performance of the Smaller UK Airports[1]

Introduction

The received wisdom is that airports, because of large fixed costs, require large volumes of traffic before they can become financial viable.[2] And, because it can take years, sometimes many years, to build sufficient traffic for profitability, it is to be expected that airports will require public sector support; indeed, it is natural for airports to be publicly owned. Of course, there comes a time when some airports have sufficient business to enable their ownership to be transferred to the private sector, but such profitable airports are 'natural monopolies' and, therefore, should be subject to economic regulation.[3]

The purpose of this chapter is to challenge this convention and to show that the industry can develop a competitive structure and within such a framework even the very smallest airport can be commercially viable. The approach is empirical using UK data. I broadly follow the Structure-Conduct-Performance paradigm of industrial organisation economics.

Industry Structure

In its 2006 Report on the UK airport market, the competition watchdog, the OFT, defined the relevant product market for its investigation as the market in the

1   Preparations for this chapter were undertaken whilst I was Visiting Professor at the Sauder School of Business, University of British Columbia in the Autumn of 2005. I am very grateful to Professor David Gillen for his support and for his comments on an earlier draft of this chapter. I would also like to thank Jan Merchant of the Centre for Regulated Industries at the University of Bath for providing in an accessible form the financial data used in this chapter and for assistance in preparing figures based on these data.

2   The argument was advanced particularly in the mid-1980s when the UK industry was subject to review prior to the 1985 White Paper on Airports Policy. See, for example, Foster (quoted in Barrett 2000), who said that economies-of-scale mean that only a few airports can be viable.

3   IATA (2007), for example, refer to the capital intensive nature of airport infrastructure making it inefficient to have more than one firm providing airport services at one location. Airports are viewed by IATA as having the characteristics of natural monopoly.

supply of airport services; these services comprised a bundled product of services to airlines, passengers and to other commercial operators using an airport. An airport, therefore, is a multi-product (or service) entity. Although airlines and their passengers are the usual focus of attention and define what is commonly regarded as an 'airport', in actual fact airports (or, in their more embryonic form, airfields) can have a much wider commercial role including shipping air freight (including mail), providing for air taxi services and general aviation, acting as a base for flying training, aircraft maintenance, flight testing and corporate jet activity, and providing for a large number of other specialist aviation services. And, even within the usual context of providing services to airlines and their passengers, there are a number different types of air service to be provided for: whole-plane charter, low-cost/no frills service, short-haul full-service carriers, and long-haul carriers; each with its different service requirements. Thus, the product market for the supply of airport services is complex and, on occasions, highly complex.

Further complexity is added because the activities of the airport company can extend beyond the supply of airport services *per se*. The property assets of an airport company might also serve non-aviation related activities. At the smaller airfields it is not unusual to find former hangers and other buildings, used for storage or as units for light industry, Gloucestershire airport provides an example of this diversification (See Box 14.1).[4] At a much larger scale, BAA plc, in 1988 acquired the property company, Lynton, with the object of exploiting its airport land holdings and agreements have been reached, for example, with the hotel industry for developments to take place on airport land.

Box 14.1   The diverse activities of small airports

Gloucestershire Airport is a successful general aviation airport. It claims to support the operations of 40 separate businesses, which directly employ over 350 people. In addition, the airport's property portfolio includes a business park which is the location of another 40 businesses employing over 1,000.

The Airport's five year development plan proposes to remove major obstacles from both ends of the main runway and to install an Instrument landing system (ILS), providing a precision approach aid for both lateral and descent paths, thus enabling aircraft to land more accurately and safely.

The current runway length is 1,421 metres, too short except for use by short-haul aircraft such as the BAe 146/Avro series, Bombardier Q400 series and the new Embraer 170. There are no scheduled services at the present time (summer 2007) in spite of which the airport currently handles around 90,000 aircraft movements a year.

---

4   Companies with property interests or broader transport interests as well as private-equity infrastructure funds have diversified into the airport industry.

It will be apparent from the foregoing that to measure the size of an airport by particular reference to passenger numbers (a common occurrence) is inappropriate. Looked at in business terms, the more appropriate measure of industry or firm size is turnover. Table 14.1 lists all those airports in Great Britain that had, at the beginning of 2006, a turnover in excess of £1 mn, 55 in total including 10 airports that form the Highlands and Islands group controlled by the Scottish Executive.[5] The airport with the largest annual turnover was, of course, Heathrow (almost £1.2 bn in 2005/6), followed by Gatwick (£361 mn), Manchester (£219 mn) and Stansted (£177 mn). These four 'designated' airports (subject to price controls) accounted for about 70 per cent of the total annual turnover of UK airports, reported to be £2.8bn in 2005 (see f.n.:16 in OFT 2006). Individual turnover figures are not available for most of the smallest airports listed in Table 14.1 but it is probable that these lie in a range between £1 mn and £5 mn; the smaller airport for which a figure is readily available is Southend with a turnover of just under £5 mn.

At the beginning of 2006, 11 of the airports in Table 14.1 were concerned exlusively with business aviation, general aviation, private flying, or related activities (although some of the remainder, such as Oxford, Cambridge and Shoreham, were served only by a single route flown by aircraft of less than 20 seats). On the other hand, there were some airfields that did not hold a formal Permission at the beginning of 2006 but, nevertheless, operated scheduled services at some stage during the year.[6] Overall, there are about 50 separate runway facilities in Great Britain providing scheduled air services.

In addition, there are more than 100 other facilities, often very small, used exclusively for general aviation and for private flying. Of these, some of the larger ones, together with aircraft manufacturing establishments with runways and former and existing military facilities, provide opportunities for future entry into the scheduled passenger market. Recent, significant, examples of military facilities converted to civil use are Newquay in Cornwall, Manston in Kent and Doncaster 'Robin Hood' airport (see Box 14.2). Previous examples include Stansted and Durham Tees Valley (both in the 1960s). Belfast City Airport, on the other hand is an aircraft manufacturing facility that introduced commercial flights in 1983 (although a more recent attempt to use the BAe facility at Filton, Bristol failed on planning grounds). Occasionally, entry has also been achieved by constructing a new airport such as East Midlands in the 1960s and London City Airport in the 1980s. But the widespread availability of existing runway infrastructure (together with increasing planning and environmental constraints) generally reduces the scope for new construction on greenfield sites.

---

5   Under the 1986 Airports Act all airport companies in Great Britain with a turnover in excess of £1 mn per annum (in two of the last three financial years) have to apply for permission to levy airport charges. Airports in Northern Ireland are regulated under the Airports (Northern Ireland) Order 1994.

6   Lydd in Kent, Newquay in Cornwall, Doncaster in South Yorkshire, and Dundee in Scotland, are examples.

**Table 14.1   Airports holding a Permission to levy airport charges (as at
1 January 2006)**

Aberdeen	Farnborough	Norwich
Biggin Hill	Filton	Nottingham East
Birmingham	Gatwick*	Midlands
Blackbushe	Glasgow	Oxford
Blackpool	Gloucestershire	Plymouth
Bournemouth	Heathrow*	Prestwick
Bristol	Highlands and Islands	Retford (Gamston)
Cambridge	Airports**	Shoreham
Cardiff	Humberside	Southampton
Coventry	Leeds Bradford	Southend
Denham	Liverpool	Stansted*
Durham Tees Valley	London City	Sywell
Edinburgh	Luton	Thruxton
Elstree	Manchester*	White Waltham
Exeter	Manston	Wolverhampton
Fairoaks	Newcastle	Wycombe Air Park

Notes

*   Airports designated by the Secretary of State.
**   The Highlands and Islands Airports are Barra, Benbecula, Campbeltown, Inverness,
     Islay, Kirkwall, Stornoway, Sumburgh, Tiree and Wick.

Source: CAA.

Conduct

The traditional relationship between airport and user is to focus on a posted tariff
of charges the most important of which are generally structured to reflect aircraft
weight. The interesting feature of this traditional approach is its informality:
the user does not need a specific contract with the airport but in paying the
published tariff the user also accepts the associated 'conditions of use' (Condie
2004; Graham 2001). The airport, therefore, assumes the long-term traffic risk.
This mattered less to airport owners when air services were subject to general
regulatory controls on route entry and thus operated in a less competitive, stable,
environment. But liberalisation of aviation has increased the risks of airport assets
being stranded by possible opportunistic behaviour of airlines that are now free
to change routes and switch airports at will. Consequently, there is an increasing
trend to establish between the airport and its downstream airline customers,
negotiated long-term contracts that achieve a better balance of risks, contracts
that are basically similar to those existing in other industrial sectors faced with
similar economic circumstances; the shipping and ports industry would be such an

---

Box 14.2   Robin Hood Airport: an example of recent entry

The new civil airport, which commenced commercial operations in April 2005, is a conversion from a former Royal Air Force airfield, which had opened in 1936. During the Cold War it was the base for part of the V-Bomber Force. It was declared surplus to Ministry of Defence requirements in March 1995 and was purchased in 1999 by Peel Holdings. Subsequently, the company invested £80 million in a new Terminal and related facilities.

During its first two years it handled over 1.75 million passengers. In March 2007 the top five routes were to Dublin, Tenerife, Malaga, Alicante and Prague. It has a very long 2,891 metre runway and Trans-Atlantic flights commenced in the summer of 2007.

The airport is located in an area from which both Manchester and Nottingham East Midland airports have drawn passenger traffic and after the new airport opened Nottingham East Midlands saw a decline in passenger numbers.

---

example.[7] In Continental Europe, this trend has extended to both full-service and low-cost airlines investing in airport infrastructure (Fuhr and Beckers 2006).

Besides specifying charges, the negotiated contract usually covers quality of service issues such as minimum expected turn-round times, airport marketing support and a commitment to future investment by the airport, together with a commitment on the part of the airline to continue to base aircraft at that airport and to develop its route network to and from it. The average charge in these, normally long-term, contracts are usually at a substantial discount to what one would normally expected to pay if use was made of the published tariff and are structured in such a way that traffic risks are shared, for example by using a per passenger charge only. The published tariff is, of course, still used for charging those airlines for which a negotiated contract is less suitable or inappropriate. But, even in this context, there has been a growing trend towards offering (sometimes for up to five years) new-route development discounts and marketing packages[8] and more clearly specified service standards, the latter further encouraged at the 'designated' airports by regulatory intervention.

The effect of the negotiated contract has been to shift the nature of competition between airports. Although airports still compete to attract linking air services provided by airlines based at other airports, the prime focus of competition is now to attract airlines (and other forms of aviation activity) to establish a local operating base at which an airline would normally position aircraft overnight and to develop from this base, or hub, a route network. Competition at this level can

---

7   For a review of similar arrangements in the electricity supply industry, see Littlechild (2007).

8   Manchester airport, for example, was intending to spend £71 mn on various forms of marketing support during the period 2003/04 to 2007/08 (see Competition Commission 2002b, Appendix 7.5).

take place over a very wide geographic market which reflects, in particular, the willingness of the low cost airlines to open new bases throughout Europe. Once a base has been established, the airline will have sunk a certain amount of costs but it will be protected from opportunistic behaviour by the airport through its long-term contract. Equally, the airport will continue to compete against rivals for the basing of future increments of airline capacity. On the other hand, those airlines that established an operating base in the past, at a time when negotiated contracts were not available and now face switching costs, are perhaps more vulnerable to increases in posted charges or other forms of opportunistic behaviour on the part of the airport. But the evidence indicates that even supposedly dominant airports, such as Manchester, are keen to see long-established airlines expand services and have been willing to discount published tariffs to achieve this.[9]

The attractiveness of an airport to an airline and thus how much an airline is prepared to pay to use it (as an operating base or as an end point on its network), depends on the quality of service factors already mentioned, and *inter alia*: the airports infrastructure (its runway length, the standard ('category') of its Instrument landing system (ILS), terminal facilities), how much spare capacity the airport has, its potential for expansion, its freedom from operating restrictions, the presence of potential competitors at the airport or a nearby airport, and, of course, upon its perceived attractiveness to potential passengers. This latter, in turn, depends foremost upon the airports location in relation to a market demand, the extent and depth of which is determined by factors such as population density, income levels, business activity, international trade links, tourism potential and the quality of the regional road network.

The CAA has suggested that a significant number of leisure passengers are, in general, willing to tolerate a journey time of around 2 hours to reach a chosen airport (CAA 2006, 22.17), although, for business travel, one hour is thought to be more appropriate. Its analysis indicates that a two-hour drive time accounts for around 80 to 90 per cent of passengers using an airport.[10] These statistics are derived from data for the larger (and more leisure oriented) airports in the UK each of which serve a large number of destinations. Consequently, they might draw from a larger than average catchment area. On the other hand, airports with a smaller volume of passenger traffic might draw most of it from a more localised area, perhaps within 1 to 1.5 hours drive time of the airport. Nevertheless, the potential for competition between UK airports would appear to be considerable as Table 14.2 indicates.

---

9   In 2002/03, for example, Manchester airport spent £1.55 mn supporting airlines/ carriers that would have otherwise have ceased or reduced services. (See Competition Commission 2002b, Appendix 7.5).

10   For Stansted the figure was about 80 per cent but closer to 90 per cent in the case of Luton and Gatwick. This difference between Stansted and the two other airports might reflect the fact that Stansted is dominated by Ryanair which has a lower average fare yield than the low cost airlines that are relatively more important at Gatwick and Luton; thus passengers might be driving longer distances to benefit from lower fares.

Table 14.2 shows driving times between proximate airports in England and Wales. Included in the data base are all those airports with scheduled passenger services and with more than 400,000 passengers in 2005/6, a total of 21. The entries in the table show times between those airports that are within 2 hours drive of each other (unless the nearest neighbouring airport exceeded 2 hours drive time). For this purpose, driving times are taken from the RAC's Route Planner and are based on assumed speeds of 60mph for motorways and 30mph for all other roads; times, therefore, are derived from very conservative speed estimates.

In spite of this conservative estimate of travel times, many airports are in surprisingly close proximity to at least one other airport (although in south east England as well as in Scotland most proximate airports are owned by BAA); bear in mind that a driving time of up to 2 hours between two airports, for example, implies that residents located halfway (in terms of time) can get by car to either airport within one hour. There is, in fact, only one airport, Norwich, lying more than 2 hours from its closest neighbour; all of the remaining 20 airports lie within 1.5 hours of at least one other airport (and, in some cases, several airports), implying a journey time of less than 45 minutes for a resident located at the half way point on the fastest route. The significance of this is that, as previously pointed out (Starkie 2002), airports (and airlines) cannot segment their customer base by residential/business location so that even a small degree of catchment area overlap can have an important effect on pricing behaviour.[11]

There have been marked differences between regional airports in passenger growth rates, and in a number of cases this appears to be associated with the establishment of new airline operating bases, which suggests that passengers do consider airport alternatives. The rapid rise during the last decade of traffic at Liverpool airport and the corresponding decline in charter traffic at nearby Manchester airport supports this supposition. In the early summer of 2006, for example, easyJet had 15 routes from Liverpool, 13 of which experienced parallel competition from other airlines operating out of Manchester, and Ryanair had 20 routes from Liverpool, eight of which experienced parallel competition from Manchester services.[12] Probably reflecting this competition, Manchester airport's share of UK air traffic has changed little in recent years in spite of a strong general growth in market share at airports outside London and the South East, the latter region having lost market share.

The competitive nature of the industry is also evident from the position taken by airport managements at planning enquiries into expansion or development of nearby facilities. Recently, Birmingham airport management objected to development plans at Coventry airport whilst management at Manchester airport

---

11   On the other hand, an airport could discriminate between sub-markets by using, for example, weight related landing charges, particularly in circumstances where not all airports have runways capable of providing for long-haul flights. Although not all the airports listed in Table 14.2 are capable of providing for intercontinental travel, nevertheless a large number can do so and the industry is potentially competitive in this segment of the market.

12   Analysis of June 2006 *Official Airline Guide*, Dunstable, Bedfordshire.

Table 14.2 Driving times between adjacent airports (hours.minutes)

	BHX	BLK	BOH	BRS	CWL	DSA	EMA	EXT	HUY	LBA	LCY
BHX							0.48				
BLK										1.44	
BOH											
BRS					1.23			1.17			
CWL				1.23							
DSA							1.22		0.48	1.20	
EMA	0.48					1.22					
EXT				1.17							
HUY						0.48				1.32	
LBA		1.44				1.20			1.32		
LCY											
LGW											1.01
LHR											0.44
LPL		1.14									
LTN	1.26										
MAN	1.34	1.01				1.44				1.06	
MME										1.29	
NCL											
NWI											
SOU			0.42								
STN											0.47

	LGW	LHR	LPL	LTN	MAN	MME	NCL	NWI	SOU	STN
BHX				1.26	1.34					
BLK			1.14		1.01					
BOH									0.42	
BRS										
CWL										
DSA					1.44					
EMA										
EXT										

Table 14.2 cont'd

	LGW	LHR	LPL	LTN	MAN	MME	NCL	NWI	SOU	STN
HUY										
LBA					1.06	1.29				
LCY	1.01	0.44								0.47
LGW		**0.44**		1.14					**1.28**	**1.19**
LHR	**0.44**			0.40					**1.08**	**1.09**
LPL					0.44					
LTN	1.14	0.40							1.37	1.01
MAN			0.44							
MME							1.04			
NCL						1.04				
NWI										2.12
SOU	**1.28**	**1.08**		1.37						
STN	**1.19**	**1.09**		1.01				2.12		

Key

BHX:	Birmingham
BLK:	Blackpool
BOH:	Bournemouth
BRS:	Bristol
CWL:	Cardiff
DSA:	Doncaster
EMA:	Nottingham
EXT:	Exeter
HUY:	Humberside
LBA:	Leeds Bradford
LCY:	London City
LGW:	Gatwick
LHR:	Heathrow
LPL:	Liverpool
LTN:	Luton
MAN:	Manchester
MME:	Durham Teesside
NCL:	Newcastle
NWI:	Norwich
SOU:	Southampton
STN:	Stansted

Note: Figures in bold show driving times between those owned by the Ferrovial-led consortium.

lodged objections to expansion plans at Liverpool and to the development of Doncaster (Robin Hood) airport in South Yorkshire. The latter is an area from which Manchester Airport has traditionally drawn substantial traffics and the Manchester Airport Group also owns nearby Humberside and the slightly-less-nearby Nottingham East Midlands Airport (see Table 14.2). Doncaster has grown rapidly since opening at the beginning of 2005 and it is interesting to observe that Humberside and East Midland airports are the only two regional airports to have shown a decline in passenger numbers over the intervening period.

Financial Performance

With such an evidently competitive industry structure, at least outside the London region and Scotland, it is to be expected that airports will have limited market power and are more likely to be price takers, especially when it comes to attracting the new low cost carriers. But if, as received wisdom will have us believe, airports are subject to high fixed costs and thus economies of density (as well as supposed economies of scale[13]), airports with small traffic volumes, and perhaps even not-so-small airports, can be expected to be loss making and to be sustained only by the provision of widespread subsidy.

To examine whether this is the case, I have used financial statistics on UK airports compiled by the Centre for Regulated Industries at the University of Bath. These data have the great advantage that they are subject to consistent (UK) accounting standards, but, nevertheless, their use is not without its problems. First, in spite of the data being compiled on an annual basis, comparison between years is difficult because of changes in accounting standards.[14] Second, and more importantly, airports have often reported year-by-year results covering different periods of time, either: nine months, 12 months or 15 months (although the latest set of data, for 2005/6, has the virtue that all airports are reporting 12 months results, albeit that the year-end does vary). Third, different airports have different depreciation policies. Finally, there are two sets of accounts available, one based on Company House returns and the other based on returns to the CAA for regulatory purposes; the two sets are for the most part the same but there are a few differences. The following analysis focuses on Company House returns for 2005/6.

There are 27 *individual* airports reporting financial data in the latest series, ranging from Southend with a turnover of just less than £5 mn at one extreme, to Heathrow with in excess of £1 bn of annual sales.[15] But, there is a discontinuity in the size range: the four designated airports, Heathrow, Gatwick, Stansted and Manchester, are very much larger than the remaining 23 and, because of this and

---

13   See Doganis (1992), for example.
14   The most recent examples are FR17 and FR 21.
15   Also included in the series, but in aggregate form only, are the results for the Highland and Island group of airports controlled by the Scottish Executive. These are excluded from the analysis here.

the fact that the focus is the financial performance of the smaller airports, the four designated airports have been excluded from the following analysis. (In addition, because of anomalies in the data, Coventry was also excluded). This gives a range of turnover for the remaining airports of between £5 mn to £111 mn.

Pertinent data for the 22 airports are shown in Table 14.3. Listed are turnover, operating profit/loss (after allowing for depreciation), net profit/loss (after allowing for tax and interest), operating profit as a percentage of turnover and of fixed assets. These data refer to all the activities engaged in by the respective airports, including what the economic regulatory accounts refer to as non-operational activities.

The data show that, in 2005/6, of the 22 airports nearly all were profitable; only two, Blackpool, with a turnover of £6.3 mn, and Durham Tees Valley, with a turnover of £10.8 mn, made an operating loss *and* an net loss overall.[16] Cardiff, with a turnover of £22.1 mn, made an operating profit, large in relation to turnover, but an overall net loss.[17] Humberside also recorded a net loss on a more modest turnover of £10.9 mn.[18]

Although the few airports recording losses of one sort or another are among the smaller airports in the group examined, there are seven other airports falling within a similar range of turnover that made both an operating profit and a net profit. These include the smallest airport, Southend; a small turnover *per se* does not appear to be an impediment to profitability. On the other hand, the margin of profit does appear to increase with turnover, but so too does the ratio of fixed assets to turnover; consequently operating profits expressed as a percentage of fixed assets, a broad indication of the return on capital employed, do not show such a strong association with turnover (see Figure 14.1).

The performance measure in Figure 14.1 does not, in most instances, suggest excessive returns in the industry, (service sector companies in the UK made an average net return of 19.5 per cent in 2006). There are exceptions, particularly London City, which might have chosen to extract location rents reflecting its exceptional location immediately adjacent to arguably the world's premier financial centre. It recently changed ownership at a price of about 30 time's earnings (before interest, tax and depreciation). Exeter and Leeds-Bradford have also changed hands recently at similar earnings multiples, perhaps suggesting an optimistic view of the growth prospects for aviation. Generally speaking the better returns were produced by airports occupying the middle range of turnover, although the relatively small Biggin Hill was a noticeable exception. BAA's non-designated airports (i.e. Glasgow, Edinburgh, Aberdeen and Southampton) generally produced returns in excess of those made in the same year at its three designated airports, in spite of the Scottish group practicing a voluntary form of cap on charges; this could of course be the consequence of superior operating efficiency.

---

16   Blackpool's operating loss and net loss were virtually identical, recording no movement on the tax account and virtually zero movement on the interest account.

17   Cardiff's net loss is the result of an exceptionally large tax charge.

18   Humberside's net loss is the result of a large interest payment.

Table 14.3   Financial data for the smaller UK airports, 2005/6

	Turnover (£000)	Operating profit/loss (£000)	Net profit/ loss (£000)	Operating profit as % of turnover	Operating profit as % of fixed assets
Birmingham Int'l	110,963	35,477	19,458	32.0	9.9
Glasgow	82,615	25,789	15,153	31.2	10.0
Edinburgh	77,381	31,381	18,335	40.6	12.1
London Luton	77,021	12,878	5,643	16.7	13.5
Newcastle	51,360	19,072	15,309	37.1	10.9
Nottingham East Midlands	50,566	15,804	7,433	31.3	25.8
Bristol	49,619	25,344	23,465	51.1	33.7
London City	40,180	7,587	6,024	18.9	164.8
Aberdeen	33,954	10,944	8,715	32.2	11.1
Belfast Int'l	31,206	9,436	4,700	30.2	7.9
Liverpool	28,799	18,336	20,606	63.7	17.7
Cardiff Int'l	22,103	5,953	–2,188	26.9	7.8
Southampton	22,023	8,791	5,941	39.9	9.6
Leeds/Bradford	21,023	1,357	571	6.5	2.9
Exeter	17,707	1,019	32	5.8	6.1
Bournemouth	14,440	2,951	1,513	20.4	5.6
Norwich	12,089	563	71	4.7	2.3
Humberside	10,934	642	–751	5.9	2.2
Durham Tees Valley	10,834	–2,715	–1,242	–9.8	–25.1
London Biggin Hill	6,893	391	246	5.7	32.1
Blackpool	6,333	–2,953	–2,952	–46.4	–46.4
Southend	4,973	137	118	2.8	7.1

Note: There is some variability in depreciation policies which will have an effect on the figures.

Source: Centre for Regulated Industries, *Airport Statistics 2005/6*.

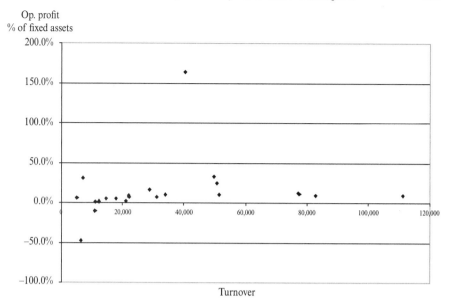

Figure 14.1 Operating profit as a per cent of fixed assets in relation to airport turnover

Conclusion

The overall conclusion is that the airport industry in England and Wales is broadly, if imperfectly, competitive and within this framework of spatial competition, small airports can be, and usually are, profitable. They appear to achieve this outcome partly by adopting incremental strategies for expansion (so that fixed costs are reasonably well adjusted to outputs)[19] and partly by spreading their activities across a large number of product lines, suggesting that for the smaller airports in particular economies of scope are important. Concomitantly, at small airports the number of air transport movements carrying fare-paying passengers on commercial flights can be a small proportion of total aircraft movements (see Figure 14.2).[20,21] But, as airports grow in size (turnover), they appear to

---

19   Although there are some instances where the length of runway or the size of other facilities seem disproportionate in relation to output, generally the assets would have been written down when ownership transferred, usually from the public sector (including the Ministry of Defence) to private owners.

20   In the case of Southend, the smallest airport in the data set, air transport movements account for only a tiny percentage of total aircraft movements so that if operating profits are expressed on a per passenger basis, Southend is more than three times as profitable as Heathrow.

21   Some of the airports listed in Table 14.1 that do not have scheduled or charter flights can have a very large number of annual movements. Gloucestershire, for example,

become more narrowly focused in terms of their range of outputs, increasingly concentrating on the higher profile activities associated with the movement of passengers on scheduled and charter flights. To this end, the larger airports appear to adopt different, more complex supply technologies (multi-storey terminals and car parks, air-bridges etc.) and, in the process, they become more capital intensive which, especially in a competitive environment, limits their returns on capital employed.

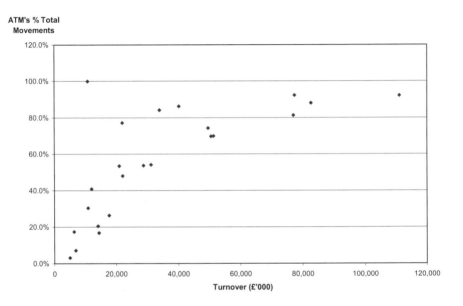

Figure 14.2 Air transport movements as a per cent of total movements in relation to airport turnover

There are a couple of developments taking place in the aviation industry that appear particularly relevant to the foregoing observations. First, with growing wealth and the globalisation of the world economy there is an increasing emphasis on private or corporate air transport, by-passing scheduled aviation with its 'public transport' characteristics. This trend appears to be taking place at a time when prospects of a technical break-through are likely to transform the economics of the very small jet aircraft, a large number of which have been ordered for air-taxi operations, in Europe as well as in North America.[22] The smaller airports are likely to gain disproportionately from such developments.

in 2005/6 had about 90,000 movements, about the same as the number of *air transport* movements at London Luton Airport.

22   The Irish owned JetBird, for example, has ordered 50 Very Light Jets and it claims to have identified up to 1000 airports in Europe that are closer to business centres than mainstream airports (*Flight International*, 15 May 2007, 43).

Second, as the influence of the low-cost carriers with their different business models becomes relatively more important, it is likely that they will attempt to seek more cost-effective solutions to the provision of terminal capacity and this may have the effect of restricting capital intensity as airports grow in size. One is perhaps seeing this played out in the current debate at Stansted and Dublin airports where there are tensions between the airport management, wishing to pursue traditional technologies in the provision of further capacity, and the low-cost airlines that are seeking much cheaper, simpler technical solutions.

Finally, the foregoing analysis begs the question: if small airports can be commercially viable, why is it that outside the UK, airports large and small are often loss-making?[23] Probably the most important explanation is that non-UK airports, in continental Europe for example, are still regarded as instruments of public policy used to promote regional development, tourism or simply to provide jobs and as a consequence still remain in public ownership[24]. But, bearing in mind the economic distortions that arise from subsidies, and the dead-weight loss associated with taxes that ultimately sustain such policies, it has to be seriously questioned whether the wider policy objectives would not be better achieved by establishing a commercially viable airport sector. This does not necessarily require a transition from public to private ownership (although in my view this would help), but it does require the promotion of competition between airports which, in turn, will require the restructuring of the industry in a number of countries. However, as Alan Walters, Mrs Thatcher's economic adviser, noted three decades ago: 'Government subsidies are a potent drug which airport authorities are most unwilling to forego' (Walters 1978).

---

23   SH&E (2006) surveyed 50 European airports and found that the average return on capital was 4.6 per cent adding that a private sector organisation would struggle with such a low return.

24   Forsyth, P., Gillen, D., and Neiemier, H.-M. (2007).

PART V
Infrastructure Access

Overview

A salient characteristic of the so-called low-cost carriers (LCCs) is that they tend to focus their activities on secondary airports, Ryanair being the exemplar of this practice. The low cost business model has been hugely successful and, in so far as the LCCs have taken business from full-service carriers, they will have helped to reduce pressures on capacity at major airports. But pressures remain and at some airports, of which Heathrow is a prime example, there is an acute shortage of capacity. This state of affairs might appear to run counter to the arguments set out in Part III of the book, namely that airports, subject to economic regulation, will be inclined to over-invest. But the contradiction is more apparent than real. It does not follow that investment is always done efficiently or is focused on where it is most needed. More importantly, there are serious planning and environmental constraints that limit both the scale and the speed at which capacity can be added, notwithstanding the ambitions of the airport owners and managers.

Consequently, there are many airports where scarce capacity has to be allocated between mutually exclusive users. The following chapters outline the mechanisms by which this is done. The most obvious way to achieve an efficient allocation between competing claims for access is to use the price mechanism, but incumbent airlines strongly oppose its use at busy airports because it reduces the scarcity rents (abnormally high returns) that they get from their access to a scarce resource at a nominal tariff (Starkie 2003a, 55; Starkie 2003b; see also Forsyth 2006). Consequently, for the most part, the adopted procedures are wanting from the viewpoint of economic efficiency. *Chapter 15* focuses upon an important exception. This exception is the trading (or secondary) market in slots that was adopted at certain US airports from 1 April 1986. The chapter is based on a study commissioned in 1991 by the Director General of Transport of the European Commission, directed by myself and assisted by my Putnam, Hayes and Bartlett colleague, Yuval Cohen. The remit was to outline the regulatory framework of the US slot market, its mechanisms and to assess the opinion of interested parties.

The purpose of the study was to inform the Commission's deliberations prior to the introduction in 1993 of slot Regulation 95/93. In spite of the study's favourable impression of the workings of the US slot market, the Commission did not embrace the idea, but in actual fact opposed it. Its opposition was driven by its intention to prize open the slot market at congested airports in order to introduce more competition; slot trading in contrast was seen to consolidate the position of incumbents who would buy slots to keep entrants out of the market and, as a consequence, further increase their dominance at congested airports. In the event, Regulation 95/93 basically adopted IATA guidelines, an administrative process centred on the incumbent's rights to slots currently used (subject to use-it or lose-it conditions), with a proportion of the slots becoming available being first offered to new entrant carriers.

Similar dominance concerns were expressed at the time by the US General Accounting Office and by the US Department of Transport in relation to the US secondary market. Both bodies noted that there had been few entrants into those airports where slots had been traded and that, after the initial shake-out when the market was first introduced, there was, subsequently, relatively little activity. But, as the study report reproduced as Chapter 15 points out, there is a different way of interpreting these events: gates at US airports are often on long-term exclusive leases and are difficult to obtain so that their availability might be the binding constraint (see also Dresner et al. 2002); or the fact that the relatively small number of trades after the initial shake-out, might suggest that the market had reached equilibrium.

A review of developments that took place in the US slots market between 1991, when the work for the European Commission was undertaken, and 2003, forms part of *Chapter 16*, first published in an Institute of Economic Affairs monograph on Airport Slots (Boyfield 2003). However since 2003 there have been further developments which have generally conspired to produce, at times, rather chaotic conditions, particularly at LaGuardia. This, in turn, has led to fresh thinking on the part of the FAA and to the advancement of further proposals in 2006. However, because no final decision had been by January 2007, the date at which the existing slot constraints were to be set aside at LaGuardia and JFK, an interim Rule was put in place in order to maintain the *status quo* pending the making of a final Rule. One is left with the overall impression that it might have been more appropriate to have stuck with the fundamentals of the original slot market, to have oiled the wheels of the market, and to have resisted the pressures of the various lobbies; but in US politics that, as they say, is a tough call.

Although Chapter 16 outlines developments in the US slot market earlier this decade, its chief purpose is, however, to provide a systematic treatment of a range of issues that have been raised by those opposing the concept of secondary markets for airport slots. These include arguments that: because of a business stealing effect and other factors, there can be a poor correlation between the amount of profit that drives opportunities for route entry and the amount of social surplus that entry creates; that airlines will engage in predatory bidding for slots to keep entrants out of the market; and, the related argument that particularly exercised the European Commission and the US DoT, that trading would result in a too great a concentration of slots in the hands of one or two airlines and would thus lead to an increase in hub dominance. Each argument opposing slot trading is considered in turn and, for each, a counterview is presented.

But, the counter to the argument that slot trading would lead to a too concentrated market is taken further in the final chapter, *Chapter 17*, which was written originally for a companion volume in Ashgate's series of aviation books:[1] it is suggested in the chapter that further concentration of slots at major hubs might have a benign effect, or even a desirable outcome, if it leads to still greater

---

1   Czerny et al. (eds) 2008.

economies associated with the natural monopoly characteristics of hub-based airline networks or to a decrease in congestion externalities (the latter effect consequent upon concentration leading to the internalisation of the externalities). Consequently, the optimal level of slot concentration is not at all clear without further investigation and this uncertainty creates a presumption against regulatory intervention, especially those interventions based on *per se* rules.

Chapter 15

Slot Trading at United States Airports[1]

The US High Density Airport Rule

Introduction

In the United States air carriers, in order not to conflict with the antitrust laws, cannot meet to discuss scheduling, routes, services or, of course, fares (although they can and do meet frequently to discuss security and gate arrangements). Partly because of this there are few restrictions at US airports limiting the allocation of landing and take-off slots; for example airline scheduling committees which have an important slot management role outside the United States do not operate at US airports.

There are access restrictions for noise-related reasons at a small number of airports including Orange County (John Wayne), Long Beach, Westchester County, Boston and Minneapolis. The restrictions usually take the form that a certain proportion of monthly operations to/from the airport have to be Stage III aircraft. In Orange County's case, there is a legal agreement that limits daily operations at the airport for noise reasons.

At San Francisco International Airport there are only 10 gates in the international terminal and this limits options for the timing of flights. As a consequence, the Chairperson of the International Terminal Operations Committee (who is the official liaison with IATA) has been given the authority to reschedule flight requests according to the availability of gates.

The major restriction in the US, however, is a formal slot allocation mechanism administered by the Federal Aviation Authority at airports subject to a High Density Airport Rule.

The High Density Airports Rule

The 'High Density Traffic Airports Rule' ('Rule') was instituted in April 1969 as a temporary measure to reduce the growing problems of congestion and delays originally at five congested airports in the US. It still remains in force at four airports and provides the framework for the US system of slot trading which was introduced subsequently. The Rule now limits the number of authorised flight operations at: Kennedy International, LaGuardia (both in New York City),

1   Report for The Director General for Transport of the Commission of the European Communities, Putnam, Hayes & Bartlett Ltd, London, 1991, published here by kind permission of the PA Consulting Group.

O'Hare International (Chicago) and Washington National.[2] Under the Rule, the Federal Aviation Administration (FAA) designated these airports as 'high density' and prescribed special air traffic rules.

The Rule established limitations on the number of Instrument Flight Rule (IFR) operations (takeoffs or landings) per hour that would be accepted at these airports during restricted hours. Each scheduled operation during the restricted hours required an air traffic control reservation, or 'slot', which permitted one operation in the same time period each day, generally for seven days a week. The total hourly limit of landing and takeoff slots for each airport was divided into three categories, one for each class of user: (a) air carriers; (b) commuter airlines (originally air taxi); and (c) 'other' operators, which consisted of primarily general aviation. Air carrier slots are required for both domestic and international flights from airports covered by the Rule. International slots are separately allocated. Finally, the hourly limits on slots were based on Engineering Performance Standards (EPS), a method for determining the operating capacity of an airport under IFRs.

Air carrier and commuter reservations are considered 'slots', which are standing reservations at the same time each day. 'Other' category reservations are allocated on an ad hoc basis for individual operations, using a first come, first served reservation system. Ad hoc reservations are available up to 48 hours in advance of the time of operation (with exceptions for weekends and holidays), by calling a voice-activated computer system.

The Rule was authorised for the purposes of efficiently managing air traffic; safety was not a factor in its adoption. The Rule was made permanent in 1973 (38 FR 29463, October 25, 1973), but remains subject to continuing FAA review and is an FAA regulation. During the course of its review process in the early 1970s the FAA adjusted the quotas at Kennedy and O'Hare airports and suspended indefinitely the controls at Newark.[3] The restricted hours and the authorised hourly limit on operations by class of user at each airport are as in Table 15.1.

Allocation of Slots Between Airlines

The High Density Rule itself did not provide a method for allocating authorised operations among the airlines. Air carrier and commuter 'scheduling committees' were established for each of the high density airports as the initial allocating mechanism. Scheduling committees were granted antitrust immunity by the Civil Aeronautics Board, (under Section 414 of the Federal Aviation Act), to discuss allocation of takeoff and landing slots.

The air carrier scheduling committee consisted of all carriers serving or seeking to serve the airport in question and available slots were allocated on the

---

2   There is a periodic slot programme implemented at Chicago's Midway Airport for general aviation.

3   Newark controls were imposed because of the gate capacity problem. When a modern terminal was built at Newark, controls were no longer seen as necessary. There is an IFR limit for Newark, but the airport is currently not subject to the Rule.

Table 15.1　High density airport rule: restricted hours and hourly/daily slot totals

	Kennedy International Airport	Washington National Airport[1]	LaGuardia Airport	O'Hare International Airport
Restricted hours	15:00–19:59	06:00–23:59[2]	06:00–23:59[2]	06:45–21:14
Air carrier slots: hourly limits	63 to 80	37	48	105 to 120
Total air carrier slots	361	670	864	1670
Commuter slots: hourly limits	10 to 15	11	14	25 to 40
Total commuter slots	62	234	252	435
Other movements	0 to 2	12	6	10

Notes

1　Washington National Airport operations are subject to modifications per section 93.124.
2　Scheduled operations at National and LaGuardia remain controlled throughout the times the airports are open for scheduled operations.

Source: High Density Airport Rule, sub-part K.

basis of a rule of unanimity – all members had to vote in favour of a proposed allocation before it could be adopted. Commuter scheduling committees, on the other hand, operated under a seniority rule whereby 'new entrants' were added to a waiting list and slots no longer needed by an incumbent carrier were given to the carrier on the waiting list for the longest period.

Using the unanimity rule for air carrier slots, each air carrier possessed a veto power over the slot allocation process. If an airline vetoed the slot allocation proposal developed by the committee, the FAA would then have the responsibility of making the allocation decision. Testimony has been offered[4] that potential new entrants (who were also members of the scheduling committee) believed that they would receive some slots if the FAA were to decide the matter, particularly after enactment of the pro-competitive Airline Deregulation Act of 1978. Incumbents, it was also argued, faced with the prospect of not knowing how many slots they

---

4　Comments of America West Airlines, Inc., before the US Department of Transportation, Federal Aviation Administration, 'High Density Traffic Airports; Slot Allocation and Transfer Methods', Docket No. 25758, February 6, 1989. See also the findings of Grether, Isaac and Plott (1981). They used experimental games based on the structure of scheduling committees.

would be forced to give up by the FAA, would also have the incentive to offer slots to potential new entrants to 'control their own fate'. Therefore, opportunities may have existed under the scheduling committee system for carriers to gain entry at the high density airports.

The evidence, however, is less convincing. New entry was 'virtually non-existent' for most of the 1970s, the period when slots were allocated by scheduling committees.[5] Slot holdings remained constant or even declined between 1972 and 1979. The key feature of the period was that there was little movement in individual carrier holdings and only one new entrant (Northeast), although new entry at high density airports was limited also by the Civil Aeronautics Board general regulation of route entry.

After deregulation, new carriers entered the industry and incumbents developed new markets. As a consequence demand for slots increased and agreement on allocations became more difficult for the participating carriers. Several scheduling committees stopped functioning and when in 1980 a new entrant, New York Air, sought to obtain a substantial number of slots to start an hourly, low-cost operation between Washington National and New York, the Department of Transportation was forced to intervene.

The Interim Operations Plan

Between 1981 and 1984, the High Density Airports Rule was superseded by the 'Interim Operations Plan' (the 'Plan'), adopted in response to the air traffic controllers' strike. The Plan established slot quotas at Kennedy, LaGuardia, O'Hare and National (as well as at many other airports) as a result of the strike's effects on the air traffic control system. It remained in operation for a total of three years, with airport-by-airport restrictions phased out as the air traffic control system was restored.

Under the Interim Operations Plan, the FAA permitted carriers to transfer and to buy and sell slots at the 22 airports included in the Plan for a six week period in 1982. During this period 194 sales took place, despite the uncertainty about how long the purchased landing rights would be valid. In addition, at least one firm, National Transportation Research Corporation, initiated a slot brokerage programme.[6] It appeared that the FAA's experiment with allowing sales of slots created a viable aftermarket in slots.

The programme for the buying and selling of slots ended in late June 1982, but slot transfers among carriers continued with the rules liberalised. For example, transfers in any ratio rather than solely one-for-one transfers, were now permitted. The FAA also adopted a lottery system to allocate additional slots becoming available at controlled airports as capacity was added to the system. Modified several times during the course of 1982, this allocation system made specific provisions for allocating additional capacity to new entrant carriers.

---

5   US Department of Transportation (1990), ch. 6.

6   Mr Roy Pulsifer, a Principal of NTRC, told us that gross turnover for his company during the six-week period exceeded $30 mn.

With the termination of the Plan in 1984, operations at Kennedy, LaGuardia, O'Hare and National were, once again, subject to the High Density Airports Rule. With the return of the Rule, the same problems encountered before the Interim Operations Plan now re-emerged. The demand for slots during restricted hours imposed stresses upon the scheduling committee system. Attempts by New York Air prior to the Interim Operations Plan to obtain slots had also changed the environment within which the scheduling committees worked. The deadlock at that time had been settled only as a result of Department of Transportation intervention: the DoT took slots from incumbent carriers for re-distribution. Several carriers challenged the legality of this move, but the US Court of Appeals upheld the Transport Secretary's decision as being within his statutory authority and consistent with the pro-competition policies of the Airline Deregulation Act. This judgment 'increased the fear of God by the airlines'.[7] The consequence of both this and the period of experimentation during the Interim Operations Plan was that by the mid-1980s there was less resistance to new approaches to the slot allocation problem.

The Buy-Sell Rule

In December 1985, Secretary of Transportation, Elizabeth Dole, signed the 'buy-sell' rule, essentially providing a new regulatory framework for the allocation and transfer of air carrier and commuter slots at High Density Rule Airports. Two features of the approach (which forms the basis of today's system) were: (a) that the initial allocation would 'grandfather' slots to the carriers holding them at the time; and (b) that a relatively unrestricted secondary or aftermarket in slots would be permitted. Effective from 1 April 1986, slots used for domestic operations could be bought and sold by any party, including non-carriers, with few restrictions.

Under the buy-sell rule:

- air carriers and commuters holding permanent slots in use on December 16, 1985, were allocated these slots, at no cost to the carrier;
- beginning 1 April 1986, any *person* was authorised to purchase, sell, trade or lease air carrier or commuter slots (except international and essential air service slots) in any number, at all High Density Rule airports;
- separate slot pools for air carrier, commuter and other operators were retained, with no change in numbers;
- slots for international operations were subject to special allocation provisions and transfer of these slots was restricted;[8]
- each slot would be given a 'priority number', assigned by lottery, that identified precisely the slot and determined its priority for withdrawal if necessary;

---

7   Interview with George Lapham, Air Transport Association of America.
8   [...].

- slots not used 65 per cent of the time in a two month period were to be returned to the FAA. That is, carriers must 'use or lose' their slots;
- a lottery procedure was established for the allocation of newly available slots and slots returned under the use-or-lose provisions. The procedure provided preferences for new entrant carriers – 25 per cent of available slots were reserved for new entrants;
- the rule stated specifically that it did not create proprietary rights in the slots;
- slots identified by the Office of the Secretary of Transportation (OST) as necessary for Essential Air Service (EAS) operations could be allocated directly to the carrier providing the service. Slots could be withdrawn for EAS operations from incumbent carriers if not enough slots were available. Once allocated or identified as an 'EAS' slot, the slot could not be transferred without OST approval;
- slots could be recalled or eliminated by the FAA for operational reasons.

Responding to criticism that the grandfathering of slots unduly favoured incumbent carriers to the detriment of those seeking entry, the Department of Transportation decided to 'kick start' the buy-sell rule by making available for reallocation to new entrants, and incumbents with less than eight slots, 5 per cent of slots at LaGuardia, Washington National and O'Hare. To make up the 5 per cent pool (some slots were already held by the FAA) slots were withdrawn from incumbent carriers by a reverse lottery. The FAA specified how many slots in each hour were required and the 'hat' was made up of 'tickets' from incumbent carriers proportional to slots held. The first incumbent carrier drawn from the hat had first choice of which slots to surrender. Slots surrendered that were not demanded subsequently were returned to carriers, in reverse order of names being drawn out of the 'hat'.

The Department of Transportation rejected arguments that a greater than 5 per cent share should be withdrawn on the grounds that service to small and medium-sized communities could be affected and that incumbent carriers had given up slots in the past to permit new entry. The DoT similarly rejected requests that periodic withdrawals and reallocations be held in the future on the grounds that: (a) new entrants *can* purchase slots; and (b) new entrants *can* obtain slots that are made available as a consequence of non-use by incumbent carriers. The DoT believed that 'once carriers begin utilising this market approach new entrants will have ample opportunity to obtain access' (51 FR 8633, 12 March 1986).

To allocate the initial 5 per cent pool of slots (and slots subsequently surrendered) lotteries were held in March and December 1986. In these lotteries new entrants could only draw up to a maximum of eight slots at each airport. Further slots which became available were distributed in lotteries in July 1987 and March 1989.

To summarise, slots are allocated using one of the following procedures:

- *Original allocation*: slots in use on 16 December 1985, when the buy-sell rule was authorised, were allocated to the carriers holding them on that date;

- *Initial lotteries*: in a one-time procedure carried-out in March 1986, air carrier slots were withdrawn from the 'grandfathered' carriers at National, LaGuardia and O'Hare to form a pool of 5 per cent of total slots at each airport. These slots were reallocated to new entrants and to some incumbents (holding fewer than eight slots) in two special lotteries held in March and December 1986. Unused slots were returned to incumbents;
- *Lottery*: when unallocated slots become available, the FAA holds a lottery for distribution of slots;
- *Administrative allocation*: the FAA allocates any slots available in off-peak hours (06:00, 22:00 and 23:00) and partial slots (slots used for fewer than five days/week) on request;
- *International operations*: slots for international operations at Kennedy and O'Hare are allocated by special procedures each summer and winter. Trading is not allowed but one-for-one exchanges are generally acceptable;
- *Essential air services*: slots identified by the Office of the Secretary of Transportation as necessary for EAS programme of contract services may be allocated directly to the carrier providing the service.

Except for international slots, EAS slots and recently allocated lottery slots, post-allocation transfer of slots is permitted on a temporary or permanent basis for any consideration. It was a requirement that carriers receiving slots through the lotteries held in 1986 should operate them for 90 days before they were allowed to sell or lease them. Since 1987 the restricted period for the resale of lottery slots has been 60 days.

Transactions, Surrenders and New Entry

Transactions

The FAA which is responsible for the administration of the buy-sell rule keeps a record of slot transactions.

These records have been analysed on behalf of America West Airlines and this analysis has been made available to us. The original analysis covered the period up to September 1988. We have been able to extend this analysis through to the end of that year by obtaining an early release of information from the FAA. The analysis concentrates on 'uneven' trades [...]; these are thought more likely than 'even', one-to-one, exchanges to involve some form of payment or recompense. The data for air carrier transactions and for commuter carrier transactions are shown, respectively, in Tables 15.2 and 15.3.

From April through to December 1986, air carrier and commuter transactions combined totalled 776, a monthly average of 86. In the following year the average number of transactions per month fell to 66, a level which was slightly improved upon during 1988. The picture for the first 2.5 years of the buy-sell rule is of an initial surge of transactions after which turnover settled at a lower, relatively stable but still high level.

Table 15.2 Monthly summary of air carrier leases and 'sales' April 1986–December 1988

M/yr of transaction	0–3 month leases	3–6 month leases	6–12 month leases	Greater than 1 year lease	Sales	Total transactions
April 1986	2	–	–	2	33	37
May 1986	1	6	–	–	22	29
June 1986	7	–	–	–	9	16
July 1986	5	2	52	–	63	122
August 1986	55	5	4	2	83	149
September 1986	22	3	16	3	81	125
October 1986	7	11	–	–	57	75
November 1986	8	5	–	–	11	24
December 1986	17	7	–	–	16	40
1986 total	**124**	**39**	**72**	**7**	**375**	**617**
January 1987	7	–	–	–	13	20
Febuary 1987	53	10	1	–	2	66
March 1987	28	1	2	–	4	35
April 1987	58	25	–	–	15	98
May 1987	35	17	–	–	11	63
June 1987	47	2	–	–	13	62
July 1987	35	5	–	–	23	63
August 1987	81	13	1	–	5	100
September 1987	43	13	–	–	54	110
October 1987	17	29	1	–	5	52
November 1987	52	8	–	–	1	61
December 1987	28	10	–	–	6	44
1987 total	**484**	**133**	**5**	**0**	**152**	**774**
January 1988	22	2	2	–	–	26
Febuary 1988	22	1	1	–	2	26
March 1988	48	14	27	–	7	96
April 1988	25	13	6	–	18	62
May 1988	18	23	2	–	12	55
June 1988	32	4	–	–	16	52
July 1988	32	11	–	–	1	44
August 1988	57	10	1	–	2	70
September 1988	45	30	–	1	2	78
October 1988	59	18	–	–	2	79
November 1988	25	12	10	–	2	49
December 1988	80	9	2	6	–	97
1988 total	**465**	**147**	**51**	**7**	**64**	**734**

Note: These calculations are based on slots operated for five or more days a week.

Source: America West, op. cit., and PHB analysis of FAA records.

Table 15.3 Monthly summary of commuter carrier leases and 'sales' April 1986–December 1988

M/yr of transaction	0–3 month leases	3–6 month leases	Over 6 mo. leases	Sales	Total transactions
April 1986	–	–	–	8	8
May 1986	–	1	–	12	13
June 1986	–	–	–	12	12
July 1986	–	2	2	4	8
August 1986	45	6	–	59	110
September 1986	–	–	–	8	8
October 1986	–	–	–	–	0
November 1986	–	–	–	–	0
December 1986	–	–	–	–	0
1986 total	**45**	**9**	**2**	**103**	**159**
January 1987	–	–	–	3	3
Febuary 1987	–	–	–	–	0
March 1987	–	–	–	–	0
April 1987	–	–	–	–	0
May 1987	3	–	–	–	3
June 1987	–	–	–	–	0
July 1987	–	–	–	–	0
August 1987	–	–	–	2	2
September 1987	–	–	–	3	3
October 1987	–	–	–	–	0
November 1987	–	–	–	8	8
December 1987	–	–	–	4	4
1987 total	**3**	**0**	**0**	**20**	**23**
January 1988	–	–	–	1	1
Febuary 1988	6	–	–	5	11
March 1988	1	–	–	28	29
April 1988	–	–	–	20	20
May 1988	10	–	–	18	28
June 1988	–	–	–	–	0
July 1988	–	–	–	–	0
August 1988	–	–	–	–	0
September 1988	–	–	–	–	0
October 1988	–	–	–	4	4
November 1988	–	–	–	6[1]	6
December 1988	–	–	–	–	0
1988 total	**17**	**0**	**0**	**82**	**99**

Notes: These calculations are based on slots operated for five or more days a week.
1   Classified as 'permanent loans'.
Source: America West, op. cit. and PHB analysis of FAA records.

However, within this general picture there were distinct differences between air carrier and commuter transactions and underlying changes in the character of the trades. In particular, there was a marked shift away from the outright sale of air carrier slots towards leasing slots for a limited period [...]. From April 1986, up to November of that year, monthly sales exceeded leases but thereafter the position was reversed. In addition, during 1987 longer term leases (those of seven months or more) were negligible. During 1988, there was a swing back to longer term leases.

For commuter carrier slots, the picture was different; sales continued to predominate over leases throughout the period. Although total transactions in this slot category were at a low level during 1987, 1988 saw a resurgence of sales such that in that year commuter slot sales exceeded sales of air carrier slots. On the other hand, trading uneven slots in the commuter market was at a comparatively low level. In proportional terms, trading in 1988 was equivalent to 10 per cent of daily slots available; in contrast the equivalent 'turnover' in the air carrier market was over 20 per cent.[9]

In addition to uneven trades, there are also 'even' trades; circumstances where carriers exchange one slot for another. These trades are less likely to include a consideration but, importantly, the buy-sell rule permits domestic exchanges to be accompanied by a consideration; (previously the FAA had instituted a requirement that carriers obtaining a slot in this way had to submit a statement that the exchange did not include a consideration). This change in the rules has made it easier to arrange an exchange such as an off-peak slot for a peak slot.

We asked the FAA to provide us with details of even trades and these are shown in Table 15.4 alongside the corresponding totals for uneven trades.

It is evident from the table that even trades far exceed the number of uneven trades and that the turnover of the former has been relatively constant. The statistics for even trades include international slots which are straight exchanges and do not come under the buy-sell rule. However, we are told by the FAA that the overwhelming majority of even trades are for domestic operations. Thus, the indications are that the overall turnover of slots in the market place is high and most likely exceeds, on an annual basis, 50 per cent.

Surrenders

Slots not utilised for at least 65 per cent of the time in a two month period are returned to the FAA for redistribution by lottery. Apart from the lotteries held in March and December 1986 to allocate the initial 5 per cent pool of slots (and slots surrendered during 1986), further lotteries were held in July 1987 and March 1989. In July 1987, there were no National airport slots available, only one LaGuardia slot (at 22:00) and only four O'Hare slots, one of which was at 06:45. In March

----

9   This annual turnover is expressed as a proportion of the 983 and 3,565 daily commuter and air carrier slots available during restricted hours at the four High Density Rule airports. In some cases, because of the short leases, the same slot will have been turned over more than once.

Table 15.4 Total slot trades 1986–1989

	Even trades[1]	Uneven[2,3]
1986 (April–31 December)	1414	776
1987	1742	797
1988	1808	833
1989	1856	Not available

Notes

1 Includes international.
2 Combines air carrier and commuter trades.
3 Based on slots operated five or more days per week.

Source: FAA and PHB analysis.

1989 no air carrier slots were available for reallocation at National, two 21:30 slots were available at LaGuardia and two 06:45 slots were available at O'Hare.

Thus there has been negligible surrender of slots to the FAA and those that became available were generally at off-peak times.

However, America West Airlines claims that carriers are manipulating the use of their slots in such a way that surrender is avoided: for the months of July/ August 1988 it compared schedules from the *Official Airline Guide* with the list of slots that should have been operated during this period. The results of this analysis indicated that there were at least 22 slots at LaGuardia and 17 slots at Washington National that were not being operated at all.[10]

New Entry

Lotteries have been used to allocate or reallocate a total of 145 'surrendered' slots (as of January 1990) to new entrants or to 'limited incumbents' (those holding less than eight slots). Of these 145 slots, only 10 are in use by the carriers that obtained the slots; most of these lottery slots have been sold on or have been taken over by an acquiring partner in a merger.

Apart from acquiring off-peak slots from the FAA or acquiring in a lottery slots taken back under the use-or-lose provisions, an airline wishing to enter the market at a High Density Rule airport must arrange for the purchase or lease of a slot from an existing carrier (or take over such a carrier). In the air carrier market very little entry has occurred as a result of slot purchase. Up to the end of 1988, America West had purchased two slots at Kennedy, whilst Midwest Express had purchased two at National and two at LaGuardia. We are not aware of any significant purchases in 1989. Thus slot purchase by new entrant air carriers appears to have been negligible.

10 See comments by America West, op. cit., 31–32.

The position in the commuter carrier market again appears to be different. We understand from the FAA that there have been a number of new entrants in this sector including Command, Business Express and Pocono; it is not uncommon, however, for such entrants subsequently to merge with or be taken over by other operators.

Views of Participants

Federal Aviation Authority

The FAA's general view is that the buy-sell rule has worked well. It has greatly relieved its administrative burden. The result is that only three or four staff members are dealing with the whole process, basically recording unequal trades and making sure that the trades filed with them are actually taking place.

The administrators did note, however, that the buy-sell rule had not met with expectations in one respect – few new entrants had purchased slots or maintained the slots allocated to them at the time of the 1986 lotteries. (Of the 145 slots allocated to new entrants only 10 slots still remain with small entrants.) Their view was that large carriers have tended to add to their slot portfolios.

They were not aware of any airline obtaining a slot under the buy-sell rule not being able to obtain a gate at the airport terminal. (In the US gates are generally leased to airlines, often for long periods of time.) It was pointed out that the regulatory framework ensured access to gates: incumbent airlines would be in breach of antitrust laws if they refused gates to new entrants and airport authorities in receipt of Federal funds were required to ensure that there was no discrimination in the use of grant-aided facilities.

The use-it-or-lose-it clause in the Rule has encouraged the leasing of slots; the incumbent benefits by having a slot used and it has encouraged better utilisation particularly of seasonal slots. But the use of leasing arrangements has meant also that since the initial period in 1986 not many slots have been surrendered as a consequence of the use-or-lose requirement.

Finally, if starting again, the administrators suggest that the system should allow for a continual reallocation of slots to new entrants. In addition, they would seek to limit 'speculation' by small carriers who, on obtaining a slot under the lottery system, sell to large incumbent airlines. A penalty provision was suggested for slots applied for and then turned-in.

Department of Transportation[11]

The major effect of serious capacity restrictions at the High Density Rule airports is that the scarce airport resource creates rents for incumbents. As a result,

---

11   The material in this section is based on the conclusions in Department of Transportation, op. cit., 2–27.

passengers pay higher fares in these markets. This scarcity rent flows from the scarcity, not from the slot allocation programme.

The slot market, while it falls short of 'perfect competition', provides an important modification to the original High Density Airports Rule, without which there would be a barrier to entry. It is clear that value exists for slots, but the price of slots contains scarcity rents and, potentially, includes rents reflecting market power. However, it is difficult to identify a market power element in the rent. There is a potential for the exercise of market power in the market for slots, and thus a potential barrier to entry. There is, however, no conclusive evidence that there are barriers to entry due to the High Density Airports Rule.

US General Accounting Office[12]

GAO was generally critical of the buy-sell rule largely because it had not allowed new entrants to start up at the High Density Rule airports. Three-quarters of trading which is taking place in air carrier slots is taking place between airlines which have some degree of cooperative agreement. The continual leasing of slots was 'effectively an absentee landlord' system.

Leasing of slots can allow the owner of the slot to ensure that the user is not a direct competitor, because the slot owner does not give up control of the slot. For example, 15 per cent of leased 'air carrier' slots, intended for the use of jet aircraft, are leased to regional carriers for use by small turboprop commuter planes. This practice prevents the slots from being used by competing jet carriers. Furthermore, such short-term leases give the lessees little assurance that long-term investment in service (e.g. by leasing gates or advertising) will be worthwhile. However, short-term leases maximise flexibility for the airline controlling a slot.

A preferred system was to establish a bidding or auction process for short-term leases with all slot rights owned by airports. More general issues mentioned involved the allocation of slots between scheduled/commuter/other categories of operator, and the fixing of the slot capacity to be subject to a review process.

Airport Operators Council International

The AOCI is an association of the governing bodies which own and operate civil airports served by scheduled air carriers throughout the world. It has over 200 members. The AOCI was basically against the buy-sell rule. It had not worked by allowing entry into the industry and it was felt that slot trading reduced the need for the FAA to improve ATC and airport facilities. It is the latter, of course, which is of chief concern to AOCI.

---

12   The material in this section is based on both interview and written material contained in US General Accounting Office, *Testimony*, Barriers to Competition in the Airline Industry, Statement of Kenneth M. Mead before the Subcommittee on Aviation, Committee on Public Works and Transportation, House of Representatives, September 1989.

The favoured approach was to abolish grandfather rights and to conduct an auction of airport slots with a certain proportion set aside for services to small communities and for small carriers. The auction would be for short leases so that there would be a continual process of re-auction as leases expired. Within this framework the AOCI would permit a secondary market to function. It suggested that proceeds from the auctions should be divided three ways: one-third to the FAA; one-third reinvested in airports; and one-third returned to airlines using off-peak slots thus establishing a peak differential to encourage peak spreading.

Air Transport Association of America

Members of the ATA (representing US carriers) at first considered the buy-sell rule 'a hare-brained idea' but after a period of operation are now more favourably disposed to it. To some extent it is Hobson's choice because the alternative – allocation by the FAA – is seen as a less desirable option.

The ATA believes that the buy-sell rule 'has worked reasonably well'. It has allowed the airlines to fine-tune their scheduling and has allowed some airlines to improve their efficiency. Compared with the scheduling committee system that preceded the buy-sell rule, the rule has allowed for a better utilisation of scarce slots; the feeling was that scheduling committees are less efficient than is the buy-sell rule.

A more sanguine view of the entry problem was expressed. Obtaining a slot was not seen as a barrier to entry because entrants could at least buy-in under the present system and it was considered unlikely that incumbents would run lossmaking services to deny entrants a slot. It was felt that the well publicised problems encountered by America West were exaggerated. The Association was not aware of many cases where airlines obtaining slots had not been able to obtain gates but, nevertheless, it was felt that airport operators could be more active in ensuring entrants did have easier access to gates.

Winthrop, Stimson, Putnam & Roberts

Winthrop, Stimson, Putnam & Roberts are the legal representatives of America West Airlines. America West is an often-quoted example of a potential entrant that has been unsuccessful in obtaining slots at Washington National and LaGuardia. The airline has testified before that FAA to this effect. Counsel, when interviewed by PHB Ltd, was critical of the existing buy-sell rule. Sales have been mostly forced sales (e.g. by Eastern) and there has been little after market. It was felt that incumbents are 'babysitting' slots, and that short-term leases are 'conveniences' for the incumbents.

Incumbents are using slots at the High Density Rule airports to serve their respective hub airports and, in this context, the value of a slot to the incumbent is much higher than the value of a slot to America West for use on a stand-alone route; incumbents when buying against a new entrant have the advantage of an existing network to build upon.

Although America West operates its own hubs, at Phoenix and Las Vegas, these lie outside the permitted distance for operations from Washington National (1,250 miles) and LaGuardia (1,500 miles). The instructive point was made that if America West could serve its hubs directly from National and LaGuardia it would be willing to pay more for a slot.

Slots should be looked at as part of a wider pro-competition policy and there was a need for a reallocation mechanism to allow new entry on a regular basis. This might be achieved by limiting entitlements to slots to a limited period; it was felt that all slots should have been auctioned initially. Additional suggestions were made that communities rather than airlines should be able to purchase slots.[13]

National Business Travel Association[14]

The NBTA was generally in favour of the buy-sell rule and it noted that the FAA had thereby established a free market system. The NBTA's main concern was that at airports where slot allocation was necessary, the available capacity was used efficiently. Generally it was thought that a competitive market would achieve this although it was recognised that financially powerful airlines could dominate a slot market. On the other hand, the NBTA was not in favour of regulating the buying and selling of slots because doing so would eliminate opportunities for competition.

Summary and Conclusions

The buy-sell rule has facilitated a large number of slot transactions at the four High Density Rule airports. These transactions fall into two groups: uneven trades when slots are bought and sold or exchanged in unequal numbers accompanied by a consideration, and even trades or one-for-one exchanges which may also be accompanied by a consideration to facilitate the exchange. Although even trades were possible prior to the introduction of the buy-sell rule, the rule, by permitting an accompanying consideration, has made one-for-one trading easier to achieve.

Following the initial surge of transactions during the last nine months of 1986, turnover in the uneven trades market settled at just under 20 per cent in both 1987 and 1988. The number of even trades has been running at twice that for uneven trades and again, on the latest figures available, shows little sign of significant decline. Total annual turnover is equivalent to more than 50 per cent of the daily commuter and air carrier slots available during restricted hours at the four High Density Rule airports.

This dynamic, fluid market in slots must be considered as a major success of the buy-sell rule. There are gains from trade regardless of whether the trades are

---

13   We note that the rules allow for any 'person' to purchase, sell, trade or lease slots.

14   The material in this section is based on correspondence with NBTA.

sales, leases or, technically, just exchanges, and regardless of whether the trades are between unrelated airlines, code sharing partners or affiliates. As was pointed out during our conversations with the ATA, trading allows airlines to fine-tune their schedules and generally improve their efficiency.

Where the buy-sell rule has failed to meet expectations and where much comment and criticism has been directed is in its apparent failure to secure new entry into the High Density Rule airports. For example, the US General Accounting Office has commented 'that the market for slots is not working as intended ... Sales of slots have declined, and potential entrants have not been able to buy slots, even when the slots were not being used by the owner. Carriers have accumulated slots in excess of their present needs and have leased out those they are not currently using rather than selling them'.[15]

Examination of the statistics does indeed show that only a relatively small proportion of transactions have been outright sales. Moreover, a number of the 64 air carrier sales that took place in 1988 were simply exchanges between acquired partners (US Air and Piedmont in March 1988), transfers between carriers and their commuter partners (three JFK slots between Pan Am and Ransome in April 1988) or transfers between carriers and cargo companies (American and Flying Tiger at O'Hare in September 1988). On the other hand, the secular decline in sales is exaggerated; a large volume of slot sales in 1986 involved the forced sale to Pan Am of 76 slots used by New York Air in its shuttle operation. (The sale was a condition imposed by the Department of Transportation when permitting Texas Air's acquisition of Eastern.) And there was the FAA's initial pool of slots distributed by lottery; (these slots were included in the statistics as sales). Nevertheless, there has been a decline in sales.

The issue is to what extent this decline and the generally small volume of sales indicates that the market has failed and that slot trading does not enable new entrants to gain access to capacity constrained airports. There are a number of points to consider.

First, carriers seeking to enter markets need more than just a 'slot'. For significant entry on one route, six to ten slots, allowing three to five rotations, are needed. Three flights a day could mean peak morning, mid-day and evening round trips, representing an intricate package. In addition, the potential entrant requires gates, counter space and handling facilities. In contrast to the general situation in Europe, facilities at US airports such as gates are leased, sometimes for long periods of time, to particular airlines. Although we received evidence that the regulatory framework ensured gate access for new entrants, it is not necessarily an easy task to secure such facilities; the DoT study concluded that 'There is very little under-utilised terminal and gate capacity at the large airports ...'.[16] It is perhaps partly for this reason that serious entry takes the form of an

15   US General Accounting Office, op. cit., 13.

16   DoT, op. cit. The DoT further concluded that 'gate facilities are a potential barrier to entry into both the aviation industry and into individual markets for firms already in the industry ... At best, the numerous contractual barriers make it difficult for a new entrant to obtain cost-competitive access to airports. At worst, contractual clauses such as [the

outright purchase of an existing operation.[17] The situation also helps to explain the apparent higher levels of entry in the commuter carrier market. Commuter carriers require less sophisticated facilities: gates are not essential and counter space and handling requirements are minimal.

Second, although sales have been small in volume and declining, this is not, by itself, *prima facie* evidence of frustrated entry. In the words of one commentator: 'it takes two to tango'.[18] We consider it important to bear in mind that the introduction of the buy-sell rule coincided with a period when the American airline industry was concentrating as a result of mergers and acquisitions. This has led, in turn, to the concentration of airline slots at the High Density Rule airports.[19] The industry has not seen aggressive new entry since 1986 and in these circumstances we would not expect to find the purchase of slots by new entrants. Consistent with this situation, slots allocated to 'new entrants' in the FAA lotteries have tended to be sold on subsequently, suggesting that there was not a market that new entrants were able or willing to exploit. With the exception of America West Airlines we are not aware of frustrated serious intent to enter the High Density Rule airports since April 1986.

Third, the developments which have taken place under the buy-sell rule, including the small volume of sales and lack of new entry, can be explained by an argument that the market has established an equilibrium; within the overall constraint of limited capacity, the most profitable opportunities are already being exploited and there is not scope for new entry. Consistent with this hypothesis is the higher level of sales in 1986 when the market was initially adjusting to the removal of a system of administrative allocation (with its potential for distortion) and its replacement by the buy-sell rule. After an initial 'shake-out' one might reasonably expect a lower level of sales and a greater emphasis on fine-tuning by leasing. The America West experience is also of relevance in this context.

America West's Counsel pointed out that incumbents at the High Density Rule airports are using slots mostly to serve their respective hub airports and that the value of a slot used in this way is higher than if used to serve a stand-alone route. Because of restrictions on permitted nonstop flight distances from Washington National and from LaGuardia, America West is unable to serve directly its own hubs in the western United States. If it was able to do so it would be willing to pay more for a slot.

Various suggestions were made during our discussions with interested parties for changing the terms of the buy-sell rule. The FAA, for example, drew our

---

majority interest clause] deter efficient development of new gate capacity, with a negative effect on new entry'.

17 Trump is an example. Trump purchased 92 slots at National and LaGuardia together with gates, as well as 17 B-727 aircraft.

18 Interview with George Lapham, ATA.

19 As an aside we note that the measure of concentration (based on the Herfindahl Index) quoted in DoT, op. cit., differs from that quoted in America West, op. cit. The latter, for the purposes of calculation, combine different airlines within the same ownership group.

attention to the possibility that the use-or-lose percentage, currently 65 per cent, might be increased. America West has suggested a 90 per cent criterion.[20]

The possibility of raising the permitted maximum seat capacity for aircraft using commuter carrier slots from the present 55 was also mentioned. A number of suggestions were also made for more radical changes. The FAA and America West felt that there should be a regular reallocation of at least some slots and the GAO, AOCI, as well as America West, favoured the idea that all slots should be made available on a short-term (leasehold) basis following an initial auction.

One suggestion we have (and not mentioned during discussions) is that trading should take place only through an intermediary. The intermediary might be a regulator, the airport operator or an independent broker(s). This would reduce opportunities for the strategic trading of slots between subsidiary airlines and other affiliates and it would increase the general level of information on the state of the market available to all airlines. The GAO has pointed out that about 75 per cent of trades are between related airlines; this is not necessarily a bad practice but trading through an intermediary is more likely to ensure that the practice is efficient.

Our overall conclusion is that slot trading has had a beneficial effect in securing a more efficient use of limited airport capacity at Washington National, LaGuardia, Kennedy and O'Hare; and that the approach is capable of further refinement and improvement. Our investigations did not reveal support for returning to the *status quo ante*, namely, the scheduling committees. However, slot trading needs to be part of a strong pro-competition policy. It does not, on its own, secure the entry of new airlines into capacity constrained airports.

---

20   The consequences of doing this need to be considered carefully. The probable effect will be to increase airline output at weekends; this raises issues of the cross-elasticity of demand between weekdays and weekends, direct demand elasticities for weekend travel and whether the dominance of certain airlines on a particular route means that there would be efficiency gains from expanding weekend output. These extensions of the debate lie beyond the terms of this study.

Chapter 16

The Economics of Secondary Markets in Airport Slots[1]

Introduction

The current downturn in world aviation markets has eased pressures on capacity at a number of major airports but in many cases capacity shortages continue to exist. There are a number of reasons for this. The overwhelming majority of airports are still publicly controlled utilities subject to political whims and tight budgets, but even where such constraints apply less (as in the UK) expansion has been hampered by environmental limitations and other planning controls. Building new runways or lengthening existing ones is not always an easy task.

As a consequence, expansion of air services has had to rely increasingly upon improvements in air traffic control technology and the adoption of new procedures to squeeze more aircraft movements through existing facilities. The ability to do so has been at times remarkable. Thirty years ago, the (Roskill) Commission investigating locations suitable for a new international airport for London adopted the working assumption that the estimated capacity of Heathrow and Gatwick combined was 440,000 annual air traffic movements: in 2001/2002 Heathrow *alone* handled 460,000. Nor has the process of improving throughput slackened in recent times. Nevertheless, the salient feature remains one of shortages of capacity; at many major airports runways are operating at, or close to, declared capacity for some or much of the day[2] and the allocation of this scarce capacity has become a major issue

Two approaches predominate for its allocation. For much of the world outside the US the basic approach is to allocate capacity according to a set of rules based on guidelines laid down by the International Air Transport Association (IATA), the airline trade association. These guidelines, first and foremost, recognise the historical use of 'slots' (the entitlement to use a runway on a particular day at a

1   First published in K. Boyfield (ed.) *A Market in Airport Slots*, The Institute of Economic Affairs, London, 2003 and reproduced here by kind permission of the Institute.

2   Declared capacity and its composition are determined by not only the runway constraints at an airport but also by the limitations of stand and terminal capacity. Even if one of the latter is the over-riding constraint this will be reflected in the capacity declaration for the runway. To simplify the argument in the rest of the chapter, we will focus on the runway constraint. For a good account of the interactions between the different components of airport capacity see Turvey (2000).

particular time, usually expressed in 15 minute blocks). An airline has a right to a slot if it has already made use of the runway at the same time during the preceding equivalent season. These entitlements, commonly known as 'grandfather rights', form the point of reference for biannual international conferences that take place to co-ordinate schedules at capacity restricted airports. At these conferences, airlines seek to modify their schedules by exchanging (transferring) between themselves their existing slot holdings, or perhaps by trying to obtain additional slots that occasionally become available (although since 1990 the IATA rules have required a proportion of available slots to be set aside for use by new entrant carriers, defined as those with negligible or non-existent presence at the airport concerned). These basic guidelines were adopted into EU law, albeit it with some minor changes, by Regulation 95/93 early in 1993. The Regulation is currently subject to review.

In contrast, in the United States, for anti-trust reasons, the IATA based system does not apply. Airlines cannot meet to discuss schedules, routes, services, nor, of course, fares. Partly because of this, there are few restrictions at US airports limiting the allocation of landing and take-off slots; for example, airline scheduling committees, which have an important slot management role outside the United States, do not operate at US airports. Airlines simply schedule their flights taking into account expected delays at the busier airports. Essentially, slots are allocated on a first come, first served basis with the length of the queue of planes waiting to land or take-off, acting to ration overall demand. Although this is the general situation in the US there is an important exception; at a small number of busy airports where the FAA prescribes the hourly number of flight operations, departure and arrival slots can be bought and sold (traded) in a secondary market. This exception is significant because it substitutes for rationing by queue or by administrative rule, a market based approach to slot allocation.

In the next section, we review the allocation of scarce airport capacity in relation to economic principles concluding that a secondary market has the potential to achieve an efficient allocation. This is followed by a consideration of a number of specific issues relating to the operations of a secondary market for airport slots. These issues include whether structural differences between markets and in the technology of supply, will lead to an inefficient allocation of slots in a secondary market, whether incumbent airlines will use a secondary market to engage in predatory bidding, whether efficiency factors can be expected to lead to a concentration of slot holdings and whether slot concentration might result in a reduction in the external costs of delay and thus produce a positive externality. The subsequent two sections consider the workings of the US slot markets and examine the hypothesis that dominant carriers might have made less efficient use of slots acquired in the secondary markets. This is followed by a penultimate section, which draws together the strands of the various arguments and also considers their public policy implications.

Optimal Allocation

Neither the approach commonly used in the US nor that commonly used in the rest of the world to allocate slots at capacity constrained airports, conforms to an economically efficient approach. The latter approach would allocate scarce capacity by basing a charge for the use of runways on the marginal cost of use. The *de minimus* marginal cost is the additional wear-and-tear caused to the runway/taxiway infrastructure by the extra aircraft movement (with other airfield costs such as fire and rescue cover, unaffected by the marginal movement). This cost is only weakly related to pavement strength and the weight of the aircraft because of other factors relating to the design of the aircraft and its landing gear; nevertheless, damage costs can be calculated with reasonable accuracy and are aircraft type specific (Hogan, Feighan and Starkie 2002). At congested airports, however, the significant marginal cost is the cost associated with the delay that a flight by one airline imposes on other airlines' flights. Delay increases fuel burn, reduces the productivity of aircraft assets and staff and is a disbenefit to passengers because increased travel time has negative value. These losses are accentuated by the uncertainty associated with the delays; both airlines and passengers schedule to allow for above average delays. To achieve an economically efficient use of slots, such externalities need to be incorporated in the user charge and added to the charge reflecting the damage caused to runway pavements by the additional aircraft movement.

Marginal cost pricing of runway use (and thus the allocation of slots by price alone) does not exist in world aviation, at least not in the precise form outlined. There are a few congested airports that make some distinction between charges for peak and off-peak use (in the form of surcharges or discounts on a standard charge) but these are typically superimposed on an aircraft weight-based tariff structure. The absence of a more efficient charging structure is partly because of the complexities of calculating a robust charge based on marginal costs. But the chief reason is the conservatism of airport authorities and especially the strong opposition of airlines to the idea of using the price mechanism for allocation purposes. At busy airports used by many different airlines, the effect of marginal cost pricing would be to increase peak charges, probably substantially, and thus reduce the scarcity rents or abnormal returns that incumbent airlines using runways at peak times obtain from their access to a scarce resource at a nominal tariff.[3] It would also have the effect of transferring the scarcity rents to airport companies leading in turn to issues of whether, for these companies, returns on capital were excessive.

There is, however, an approach, which has the potential to square the circle: incumbent airlines retain their rights to slots already held under the grandfather rule but are allowed to sell these rights to non-incumbents (or to buy further slots themselves) in a secondary market. In the process, high added-value users of runway capacity compensate, through the purchase price, users with low added

---

3   At off-peak times, marginal cost based tariffs would probably *decrease* from current (average cost) levels.

value so that slots are transferred to airlines best able to add value to the service network. By allowing the market to redistribute slots to services from which airline passengers derive most benefit there is a net gain in welfare.[4] (Although, because scarce capacity is now utilised by flights from which passengers derive most benefit, it is also possible that the overall average fare yield at congested airports will increase as a consequence.) The secondary market thus provides an opportunity for an efficient allocation of slots to evolve, removing in the process the allocative inefficiencies resulting from price controls and administrative rationing.

The introduction of a secondary market in airline slots will mean, of course, that the incumbent slot holder receives a lump sum financial benefit when a slot is purchased either by another incumbent or by an entrant and this financial transfer might be considered to give incumbents an undue advantage. But this lump sum represents the present value of the scarcity rent associated with the slot. The incumbent is currently receiving this scarcity value through the yield premium on fares charged to passengers.[5] In this respect those with grandfather rights already enjoy the 'windfall' and the introduction of a formal slot market will neither add to nor subtract from this.[6]

The Potential for Misallocating Slots

The argument that a secondary market will secure an efficient allocation of slots assumes that airlines competing for slots are also competing in the market for passengers. Where this is not the case, where, for example, bilateral air service agreements restrict the frequency of service and number of carriers on a route, it could be argued that an optimal allocation of slots would not be the outcome of a secondary market. This is because market expansion by the route monopolist would lead to a fall in yields as lower prices are paid by existing as well as new traffic. An airline expanding to the same degree in a more competitive and thus price elastic market would see yields eroded less. Consequently, the monopolist would be less interested in purchasing slots in a secondary market and might, instead, become a net seller of slots. The result could be too few slots used (and services provided) in less competitive markets relative to the more competitive markets.

In addition, it has also been argued that even when there are no administrative restrictions on market entry, for example in intra-EU markets or in the US domestic market, a secondary market for slots will not necessarily secure their

---

4   A legitimate question is: why do not those airlines with historic rights serving low value markets switch services to more highly valued markets? In other words, is not the division of the market portrayed here, rather artificial? The division reflects both barriers to entry associated with international air service agreements and the absence of a market in slots; airlines do not have an incentive to earn a return on assets that are freely acquired and are not fully tradable.

5   Recent analysis by the CAA (2002, 14–18) suggests that for an airport like Heathrow the yield premium can be quite large.

6   See Starkie and Thompson (1985b) for elaboration.

most efficient use (Borenstein 1988). Borenstein has argued that the use of a slot in any particular market is driven by the opportunities for profit, but there can be a poor correlation between the amount of profit and the amount of social surplus and that this difference will vary from route to route. The difference will be driven by the use of different production technologies by different firms (airlines), by the varying attributes of different markets in terms of demand elasticities, by the number of incumbent firms and by the opportunities for price discrimination.

In particular, it is argued that the equality of profits and social surplus, and hence equality between profits and efficiency, fails for two particular reasons. First, lumpiness in supply means that infra-marginal output produces consumer surplus that cannot be captured by the firm. And, second, entry onto routes with incumbent airlines will have the effect of transferring some rent from the incumbent(s) to the entrant; this transfer does not change the overall social surplus but, nevertheless, it accrues as profits to the entrant. This will cause the entrant to over-value the licence (slot) and possibly lead to too much entry in crowded markets, and vice versa.

When Borenstein came to consider the airport slot market (his initial arguments were couched in the general context of licences to operate in restricted markets) he was, however, more sanguine. This was because the amount of capacity that could be provided by any single flight was limited, so that he acknowledged that capacity additions might be more welfare enhancing in established airline markets than his earlier abstract analysis had supposed. Moreover, since Borenstein wrote his paper, technical developments have reduced the scale of the problems he raised. The introduction into service of regional jets has provided a technology that allows for smaller increments of supply and improvements to and the spread of yield management techniques have enabled firms to finely tune prices in most aviation markets. These developments, in turn, increase the ability of all airlines to extract more surplus as profit and, therefore, the variation in demand elasticities across markets is likely to be less of a problem than Borenstein originally supposed.

By the same token, the ability that now exists to price discriminate within the market in a sophisticated manner also reduces the likelihood that the route monopolist will restrict supply and thus be inclined to purchase fewer slots in a secondary market.[7] The importance of price discrimination is a point that we return to later in the chapter.

---

7   If this is considered to remain a significant problem it might be addressed in several ways. First, slots used on routes where entry is restricted by, for example, bilateral agreements could be ring-fenced and excluded from the processes of the secondary market thus preventing the transfer of slots from less to more competitive markets. This is, in effect, the approach used in the US trading market where trades are limited to slots used for domestic services. Second, additional slots that result from improvements in runway utilisation, or air traffic control procedures, could be restricted for use only on routes operated in a less competitive environment. Third, a differential tax with a higher rate applying to competitive routes could be introduced. The tax difference would equate to the mark up of fares over airline operating costs on the monopoly routes. However, the information requirements for such an approach are demanding and air service agreements would probably prevent its implementation.

Trading, Predation and Anti-competitive Behaviour

A further argument, advanced in opposition to the view that a secondary market achieves an efficient distribution of slots, is that airlines will engage in predatory bidding for slots. Specifically, the argument is that established airlines with grandfather rights might buy slots to keep entrants out of the market and, as a consequence, further increase their dominance at congested airports. This argument has been subject to close examination by McGowan and Seabright (1989). They accept that it has substance but are of the view that, for established airlines, it is an expensive way to deter or drive out competitors. This is because at any one airport there are many slots, each one of which has a large number of close or reasonably close substitutes (bearing in mind that slots are transferable between services). To keep a newcomer out of a particular market, an incumbent airline might, therefore, have to 'overbid' on a large number of slots. In these circumstances, it is argued that it is more likely that an established airline will direct any predatory behaviour to the route (service) itself where it will be more focussed.

At an airport like Heathrow, where the largest airline (British Airways) and its alliance partners holds around 40 per cent of the slots overall and a similar proportion in the peak, this argument appears reasonable: attempts to keep potential entrants out of the market by overbidding on slots will be expensive and is unlikely to occur. But the McGowan and Seabright counter argument is weaker at an airport such as Frankfurt, where over 60 per cent of slots are in the hands of Lufthansa, or Atlanta where Delta's share exceeds 70 per cent. Here the number of slots on which these incumbents would need to overbid could be small. Consequently, at an airport where the incumbent already holds the majority of slots, overbidding might be a feasible way of preventing entry.[8]

There is a need, however, to take an additional factor into account. Once a formal secondary slot market is introduced at an airport, its slots become tradable assets and, as a result, their value should be written into the balance sheet of the airlines using the airport. This book value will reflect the market value of the slot, which in turn will reflect the capitalisation of the economic rent that the average airline can command from the use of a slot. Consequently, slots will be bought when an airline is able to earn a satisfactory return on the investment (because the purchaser judges that it will be able to extract more rents from the acquired slots than the airline selling the slots) and slots will be sold when their worth to an airline is less than their market (book) value.

In this context, over-bidding the price of slots to remove competitors from the market, or to keep entrants out, could result in the acquisition of poorly performing assets, although this has to be balanced by the higher returns on

---

8   Wolf (1999) suggests that an airline could try to monopolise the slots by making a one-shot offer to current slot holders to buy all slots at a fixed and *ex ante* pre-specified price, with the offer terminating if there is not a complete acceptance. Of course, such a strategy would be conspicuous to the competition authorities and would probably fail on competition grounds.

other slots now subject to less competition. Where the balance will lie is not self-evident, but the complexity of the trade-offs involved suggests that predatory behaviour, should it take place, is more likely to focus on the level of fares set for routes subject to, or threatened with, entry.

Incentives for Slot Concentration

Although the concerns of the competition authorities have been driven by the prospect that trading could re-enforce the position of large incumbent airline at major airports, there are in fact a number of reasons why efficiency considerations might lead to a concentration of slots and to an increase in dominance at such locations. In large measure, incentives to concentrate slot holdings stem from the well-known economies associated with networks and the advantages of concentrating flows in networks on particular nodes referred to as hubs. The number of potential connections at a hub airport increases exponentially with the increase in the number of markets served from a hub; as a consequence there is a positive externality generated by concentrating flights. Potential travellers are faced with increasing opportunities to fly to and from their preferred end points, which increases the attractiveness of a hub airport as more and more cities are added to the network. In addition to these attractions of the scope of a hub network, the concentration of traffic at the hub allows for increased service frequencies along routes radiating from the hub. This increase in frequency enables passengers to match more closely their preferred time of travel with the scheduled time of the flight and thus to reduce schedule delay. These advantages to the passenger are not without some associated costs in the form of more circuitous and longer journeys but, overall, the passenger is argued to gain from airlines concentrating their flights by making an airport a hub (Morrison and Winston 1986).

In addition to these demand-side benefits of network scope and density, there are potential supply-side gains from concentration. Concentration can generate cost efficiencies because of economies of route density (with fixed station costs, for example, spread over more units of output as a network of pre-determined size is used more intensely) and because of cost economies of network scale and scope (with larger networks enabling the better utilisation of aircraft fleets for example).[9] The result of these scale economies, together with the demand synergies of networks, is that a slot can be more valuable if it is used to enhance an existing network of services.

These demand side economies which encourage airlines to focus their operations at particular airports, do not in themselves predicate hub dominance. It is to be noted that the demand-side economics of the hub system could work equally well if every end point on the network was served from the hub by a

---

9   Ng and Seabright (2001) using a panel of 12 European and seven major US airlines for the period of 1982 to 1995 find returns to density of 1.19 and returns to scale of 1.09, although the indications were that returns to scale were exhausted at higher levels of output.

different airline. Similarly, that part of the supply-side gain that comes from the economies of density could also in theory be realised if each route was served by a different [but singular] airline. This is not true, however, in relation to the economies of network scale and scope: in a concentrated network these benefits accrue from airline concentration. In addition, studies have also shown that passengers have a preference for making on-line connections rather than transferring between airlines even where such transfers are formalised through IATA interline agreements.[10] This preference reinforces the attractiveness of the airline that happens to have the larger presence at the hub. In these circumstances and in the absence of regulatory constraints, at major airports with limited capacity one might expect both consumer preferences and cost efficiencies to lead to an increasing concentration of slot holdings.[11]

Congestion Externalities

The traditional approach to the measurement of congestion externalities assumes that the delay that one flight imposes on all other flights is a true measure of the external costs imposed by that flight. But recent economics literature on the costs of airport congestion has argued that this approach overstates marginal costs (Brueckner 2002; Mayer and Sinai 2002). It does not take into account the fact that a particular flight might very well impose delay on other flights operated by the *same* airline and that the delay costs that an airline imposes upon itself are, in effect, internalised and no longer constitute an externality. In these circumstances, treating the delay that each flight imposes on *all* other flights as a cost is only valid for airports where each and every flight is operated by a different airline. At the other extreme, an airport totally dominated by one airline cannot experience congestion externalities: congestion might exist but its level is of the monopolist's own choosing.

This argument has particularly relevance at those airports chosen as hubs and at which dominant airlines concentrate their schedules into limited times of the day. At airline-dominant hub airports, the hub airline will seek to organise flight schedules into banks or waves of arriving flights mirrored by waves of departing flights with the intention of minimising passenger transfer times. Such concentration of flights into narrow windows of time increases the likelihood of congestion, as does the process of adding more end-points to the network. But the passenger is gaining from reduced connection times and from the increasing reach of the network so that longer delays are the outcome of a hub airline equating the high marginal benefits due to increasing concentration, with the increasing marginal cost of delay; a decision process which internalises the externality in seeking to reach an efficient equilibrium (Mayer and Sinai 2002).

One implication that follows from this view, that not all delays constitute an external cost, is that an optimal congestion charge will vary depending not only on

---

10   See Economics-Plus and GRA (2000) for an analysis of this effect. [...].
11   These same factors encourage airline concentration (see Bailey and Liu 1995).

the amount of delay that each additional flight imposes, but also on the proportion of flights at an airport operated by each airline. Specifically, the appropriate charge is the delay cost imposed by the additional flight multiplied by one minus each airline's share of total slot numbers (Brueckner 2002). Consequently, an additional flight by one airline with 50 per cent of the movements at an airport would pay half the level of (congestion) charge paid by an equivalent flight operated by a new entrant airline.[12] Equally, if the largest airline operating at a congested airport increased its slot holding from, say, 50 to 55 per cent of the total, the congestion charge paid by its flights should fall by 10 per cent, enabling the airline to capture the internalisation of the delay externality in reduced charges. Thus, if slot concentration reduces the delay externality this benefit from slot concentration should be reflected in prices charged. Of course, as we have noted, efficient pricing of delay is difficult to achieve (hence the current focus on secondary markets). Thus, given the constraints on setting optimal prices, the appropriate response is to pay a Pigouvian subsidy; the airline acquiring slots in the secondary market should receive a subsidy equivalent to the reduction in the delay externality associated with its acquisition.

Market Power

Although increasing the concentration of slots at congested airports reduces the delay externality, this comes at a price: increasing concentration potentially increases market power. Whilst the passenger might benefit from the optimisation of flight schedules that concentration allows for, this could be at the expense of fare increases unrelated to the underlying costs of the hub system. In the US, for example, although airlines compete across their respective hubs so that many passengers have a choice of flights via competing hubs, such choice does not extend to those passengers starting or terminating their journeys at a hub airport. US studies have shown that such passengers appear to pay fare premia, although it is arguable whether these premia reflects scarcity rents and the extraction of consumer surplus by yield management (price discrimination) techniques.

Such is the sophistication of price discrimination methods using airline CRS systems that some have argued that the outcome in a monopolistic market is a level of output (and thus a combined level of consumer and producer surplus) similar to that which would prevail in a competitive market (Button 2002). Because airlines have the ability to segment the market and price-discriminate between individuals or small groups of passengers in a highly effective way, the effect is to remove in large measure the difference between the (declining) marginal revenue curve and the average revenue curve. In seeking to fill capacity, airlines sell seats at many different prices shifting down the demand function as they do

---

12   Note that this outcome depends upon a stable and recurrent daily pattern of flights. If, in contrast, an airline accounted for 90 per cent of peak flights for half the period and 10 per cent for the other half, giving 50 per cent overall, a charge based on the latter would be too low in the former sub-period and too high in the latter sub-period.

so until the price of the marginal seat is equal to its marginal cost, which is the competitive outcome.

This suggests that a situation has evolved where concentration might have a benign impact because airlines, even airlines with market power, act in an allocatively efficient manner when pricing their product.[13] Arguably, this might appear to make competition policy redundant in the airline sector, but it does of course ignore the issue of productive efficiency: although airlines (because of incentives to fill seats) might be producing at optimal levels of output given their respective cost bases, costs might nevertheless be at inefficient levels, a possibility borne out by many studies of airline costs.[14] This suggests that the real importance of competition in the airline industry is to bear down on cost inefficiencies. To the extent that this can be achieved through capital market disciplines or, as findings by Richards (1996) and Morrison (2001) suggest, without the need for head-to-head competition, a more relaxed view of slot concentration at congested airports can be taken.[15]

The US Trading Markets

A secondary market approach is used currently at a small number of busy airports in the US as an alternative to rationing by queue. Because the demand for runway use was particularly high at JFK and LaGuardia (LGA) in New York City, at O'Hare (ORD) in Chicago and at Washington Ronald Reagan National (DCA), many years ago the authorities stepped in to prescribe a limit on the number of flights at these airports during specified hours. The limit was known as the High Density Rule. The Rule did not itself provide a method for allocating the authorised number of runway operations between airlines and a number of different approaches were tried without a great deal of success before trading markets were sanctioned at the end of 1985.

The market was restricted to slots used for *domestic* services which in turn were divided into two groups: air carrier slots, and commuter carrier slots (originally operated by aircraft with 56 or fewer seats). Slots assigned for use by commuter carriers could not be purchased by air carriers. In addition, slots

---

13   Note however that the dominant airline can extract monopoly rents in different forms. For example it could do so by imposing delay costs on passengers in order to realise cost savings in the use of equipment and staff. However, the extent to which the airline monopolist would wish to extract rents by increasing passenger delays are probably limited; such delays are correlated with operational delays that, in turn, increase airline costs. Clearly such increases are offset by strategic gains from scale and scope economies and from revenue enhancements that flow from adopting a hub system with a pronounced peak pattern of operations. Nevertheless, dominance provides the opportunity for inefficient behaviour.

14   See for example Ng and Seabright (2001) and Windle (1991).

15   This suggests that policies that encourage the establishment of low cost airlines and the privatisation of state airlines, have a major role to play in increasing the efficiency of the airline sector as a whole.

used for subsidised 'essential air services' were excluded from the market. The regulations stipulated that any person was entitled to purchase, to sell or to mortgage a slot, or to lease on a temporary basis and third parties such as local communities could be included in the transactions.[16] However, slots not used for a stipulated minimum of time in a two-month period had to be returned to the FAA; that is to say, carriers had to 'use-or-lose' their slot. Surrendered slots, or others becoming available, were assigned to a pool and reallocated using a lottery but with 25 per cent initially offered to new entrants.

These basic terms were introduced in April 1986 when airlines started to buy and sell those slots that they were holding as of 16 December 1985. Later, small amendments were made to the regulations. From January 1993 slots had to be used for 80 per cent of the time in a two-month period (it was previously 65 per cent) and the definition of those entitled to slots from the reserved pool was broadened. In addition, restrictions were introduced to prevent slots intended for new entrants being acquired by incumbents [...]. Other amendments adjusted the distinction between air carrier and commuter slots, a distinction that was introduced originally in order to strike a balance between maximising the economic use of runway resources and preserving services to smaller communities. The aircraft size threshold for the use of commuter slots was increased, particularly at ORD.

During the 1990s, however, the Department of Transportation was subject to constant pressure from politicians and lobbyists seeking privileged access for either small communities or for competing airlines. As a consequence, exemptions were granted at LGA; a new entrant Jet Blue obtained a significant number of slots at JFK; and there were some exemptions granted at ORD. In spite of, or perhaps because of, these concessions, pressures continued and eventually Congress passed ARI21 leading to a radical change in policy and proposals to phase-out of much of the High-Density Rule. The Rule was eliminated at ORD from July 2002. It is to be removed from JFK and LGA in 2007, although it will continue to remain in place at DCA. In the meantime, open entry has been declared for regional jets, except at DCA. This has led to a huge expansion of such flights and, in turn, to very serious delays with the result that the FAA has intervened and imposed a lottery on a temporary basis at LGA. Not surprisingly, these changes in the regulations and the 2002 phase-out of flight limits at ORD, have resulted in much reduced activity in the slot markets. At LGA the value of slots has been diluted by the various exemptions and recently, by the lottery.

Analysis of the Trading Markets

Data on air carrier slot transactions for the period from 1986, when the regulation was first introduced, to 1992, prior to the amendment of the regulation, are shown in Table 16.1. During the first six to nine months, there was an initial surge of sales as air carriers acquired the slots they believed they could use best and disposed of those that could be sold profitably. This initial phase was

---

16   For further details see Starkie (1992 and 1994b).

followed by a decline in the number of outright sales while the number of leases grew, particularly short-term leases possibly reflecting a requirement for the use of a slot at limited times of the year only. After 1988, however, sales once more increased such that in both 1990 and 1991 they exceeded the total for 1986. The number of lease transactions also increased considerably between 1988 and 1989 and thereafter stabilised at the higher level. Separate information is also available for the ring-fenced slots used by commuter carriers but only for the first three years (Table 16.2). Commuter carriers appeared at that time to be more inclined to buy and sell, rather than to lease, and some new commuter airlines took the opportunity to enter the market. A significant number of slots were held by non-carriers; some airlines mortgaged their slots to financial institutions whilst local communities also took the opportunity to purchase.

Table 16.1   Yearly summary of air carrier slot leases and sales at US high density airports 1986–1992

	Leases		Sales	Total transactions
	< 6 months	**> 6 months**		
1986	163	79	375	617
1987	617	5	152	774
1988	612	58	64	734
1989	1,259		290	1,549
1990	1,294		403	1,607
1991	1,468		477	1,945
1992	1,178		310	1,488

Source: Starkie (1992) and Wolf (1999, 126).

Table 16.2   Yearly summary of commuter carrier slot leases and sales at US high density airports 1986–1988

	< 6 months	**> 6 months**	**Sales**	**Total transactions**
1986	54	2	103	159
1987	3	0	20	23
1988	17	0	82	99

Source: Starkie (1992).

At a superficial level these, albeit limited, data appears to bear out the view of the US General Accounting Office (USGAO 1996) that there had been few entrants into the four US airports and that established airlines have increased their share of slots. There does appear to have been relatively few outright sales, particularly when account is also taken of transfers as a result of mergers and other statistical quirks (see Starkie 1992, 15). However, the evidence has to be balanced by other considerations. First, airlines when entering the market need more than just a runway slot; they also need access to gates, counter space and handling facilities. In contrast to the general situation in Europe, facilities such as gates are leased at US airports and spare capacity is not always available. Second, declining sales and lack of entry has to be interpreted in the context of developments in the airline market as a whole. Generally throughout the period the American airline industry was consolidating as a result of exits or mergers and acquisitions and in these circumstances it might be argued that entry at the High Density Rule airports was less likely to occur. Third, the emphasis on the short term leasing of slots rather than on outright sales, might be explained by the market approaching equilibrium; within the overall constraints of limited capacity, the most profitable opportunities were being exploited to a very large extent by the established airlines. Therefore, it is difficult to conclude from the general picture with limited data available, that slots were used inefficiently and perhaps hoarded.

Kleit and Kobayashi (1996) have undertaken a more rigorous analysis of whether slots have been used efficiently by using data for Chicago O'Hare. Of the four US airports where slots are bought and sold, ORD has the most concentrated holding of slots (reflecting its role as an airline hub) and it is also the one airport where those airlines established before deregulation of the US domestic market have increased the number of slots held. The analysis, based on 1990 data, focused on the utilisation of slots. It specifically examined whether the two large dominant carriers (United and American) were using their slots more, or less, intensively than the smaller carriers: the usual market power argument would be that the dominant firm(s) have an incentive to restrict output which would be reflected in lower utilisation rates. Slot utilisation was measured in three ways: by the average rate at which slots were used; by whether leased slots were used more or less intensively than owned and operated slots: and by the average daily seat capacity per slot. The analysis indicated that the dominant carriers had a higher usage rate for their slots and that slots that they leased out were used at an equal or higher rate than owned and operated slots. In other words, there was no indication that dominant carriers were hoarding poorly utilised slots, or were leasing slots to other airlines that would make less use of them. On the other hand, there was evidence that one of the two dominant carriers (United) was using on average smaller aircraft.[17] Therefore Kleit and Kobayashi were unable to confirm that their analysis unequivocally supported the argument that concentration had led to increased output from the slots concerned but the conclusion that they arrived at was that concentration in the slot market at ORD did not appear

---

17   This might be explained by differences in the markets served by dominant and non-dominant carriers. For example, holders of a smaller number of slots (e.g. Continental) might have used these for long-haul flights. Alternatively, the fleet used by United might not have been fully optimised in relation to the dynamics of the market because such adjustments are not always economic in the short term.

to be leading to anti-competitive behaviour. The evidence was considered more consistent with the hypothesis that it was efficiency considerations that were generating concentration.

Policy Prescriptions

Established airlines operating at major airports appear to gain significant benefits from expanding their slot portfolios. Expansion is encouraged by the potential for economies of scale, scope and density from operating a network of services. Fixed costs are spread over more units of output and the characteristics of a connected network drives market demand, enhancing revenues. Passengers benefit from the reach of the network, from increased flight frequencies and from increasing opportunities for making on-line transfers.

Although the busier the airport the more likely it is to experience congestion, some of this delay can be self-imposed. By adding a flight an airline potentially adds to the delay experienced by its other flights at that airport, but it takes this interaction into account when deciding upon the scale and timing of its schedules. The effect is to internalise the normal external costs associated with congestion. It follows that by increasing slot concentration, delay externalities are reduced. If airport charges were based on the marginal costs of use, including marginal costs of delay, a dominant airline adding to its slot portfolio at the expense of other airlines, would capture in a reduction in airport charges the positive externality associated with its decision. Absent marginal cost pricing of runway use, the second-best solution would be to pay a Pigouvian subsidy to the slot-acquiring incumbent airline, equivalent to the reduction in the external costs of delay. In these circumstances it is to be questioned whether the common concerns of policy makers – that large incumbent airlines might use a secondary market to acquire more slots at the expense of small incumbents or new entrants – is not misplaced.

There is a genuine dilemma here. Slot concentration is driven by efficiency considerations and there would appear to be considerable gains in the offering. But, at capacity constrained airports, if a secondary market achieves efficiencies through facilitating slot concentration, by the same token it also has the effect of reducing opportunities for service competition at the route level. Competition has then to come from alternative hub airports probably equally dominated by one or two airlines, or from secondary airports used by low-cost carriers. This competition can be less effective especially for short-haul flights and for those passengers originating or terminating their landside journeys close to the hub airport.[18] On the other hand, the sophistication of CRS-driven yield management techniques does enable airlines to capture rents (consumer surplus) through extensive price-discrimination, so that the normal adverse efficiency consequences of limited competition, the curtailment of output, might be much

18   Airport competition is a relatively neglected issue that warrants more attention if only in the light of the arguments presented in this paper. See also Starkie (2001).

less in this particular case. The situation is complicated however; airlines could exploit their dominance by discretely imposing delays on passengers, the effect of which is a loss of efficiency. Although it is argued that delays at hub airports represent an efficient equilibrium between the benefits of concentrating flights into narrow windows of time and the marginal costs of increasing delay, this is not necessarily so if the passenger is poorly informed about expected delays and published schedules are too optimistic.[19] But in such a case the remedy does not lie in imposing constraints on the use of slots.

Where should the balance lie between the undoubted benefits of concentrating slot holdings and the potential for abuse of the dominance that concentration facilitates? The answer is difficult and to an extent circumstantial. At airports like Heathrow or Kennedy, both with capacity fully utilised for much of the time, but with the largest airline(s) occupying only a modest share of the total number of slots, there would seem to be much to be gained by allowing slot concentration through the operation of a secondary market. The diminution of the congestion externality alone is a powerful argument for doing so, but there might also be large gains from improving the timing and the schedule co-ordination of the dominant airline(s). In these circumstances, to impede slot concentration by legal or administrative constraints is likely to lead to significant losses of efficiency. On the other hand, at a capacity constrained airport heavily dominated by one airline group, such as at Amsterdam Schipol, the balance is much finer; there are probably diminishing returns from further concentration, but increasing risks from a less competitive environment potentially harmful to consumers. In these circumstances with the overwhelming majority of slots in the hands of one airline, it might not be appropriate to allow the dominant airline to purchase further slots in a secondary market without showing just cause.

Conclusions

The characteristics of airline networks are such that major airports frequently develop as major locations for the transfer of passengers between flights. They become hub airports and because of capacity constraints they frequently experience congestion. At such a congested airport, the tendency will be for a single airline (or occasionally two airlines) to dominate. Passengers when transferring between flights prefer to change between aircraft belonging to the same airline; on-line transfers are generally preferred to interlining; and, since the number of potential connections increases exponentially in the number of the markets served by the hub airline, the hub dominant airline has an incentive to expand its network. This characteristic, together with potential production economies associated with the scale of the network and the attractions of higher service

---

19   An airline could, for example, publish optimistic schedules (that influence passenger demand) and use the delay imposed on passengers to complete rotations on longer haul routes, thereby obtaining better equipment utilisation.

frequencies, means that the hub network operated by a dominant airline, has many of the characteristics of a natural monopoly.

The development of the natural monopoly network can be stymied, however, by the difficulties of gaining access to runway slots which, the US apart, are allocated primarily on the basis of so-called historic rights. This would not matter if charges for the use of airport infrastructure reflected the marginal cost of use, in which case the price mechanism would allocate slots to those who derive most value from their use. But, airport charges are rarely based on marginal cost and there remains considerable opposition to such an approach. Consequently, the allocation of airport slots frequently is sub-optimal. It is a situation that can be rectified however by allowing airlines to trade slots in secondary markets. Such markets have been operating at four US airports since 1985 and the indications are that these markets have effected a more efficient distribution and use of slots. In essence, given the constraints on correct pricing signals for the use of scarce airport capacity, secondary markets for slots would appear to be an important way of satisfying the dynamics involved in airline markets and an important means for discovering the efficient scale of the network natural monopoly.

Nevertheless, a one-size-fits-all approach is not necessarily appropriate. Opportunities for the reallocation of slots allowed for by a formal trading market are likely to be of greatest value at those congested airports still used by a large number of airlines and where no single airline has a majority holding of slots. At these airports, in addition to the realisation of network economies, consolidating slot ownership is likely to reduce significantly the external costs of delay. In contrast, when most slots at a major airport are in the hands of a single carrier, both the potential gains from trade and the positive externalities from slot concentration, will be less and the potential for abuse of a dominant position greater. In these latter circumstances, a trading market is much less likely, although by no means unlikely, to lead to more efficient outcomes.

Chapter 17

A Defence of Slot Concentration at Network Hubs[1]

Introduction

A number of the worlds' major airports have a high proportion of their capacity utilised by a single airline, or alliance of airlines. At major US airports, for example, it is common for the leading carrier to account for three-quarters or more of the flights; at European airports the proportions are generally smaller but often exceed 50 per cent […]. This has lead to concerns that such high levels of capacity utilisation ('slot concentration') will impact adversely on competition and prices and, in turn, has inclined policy makers and regulatory bodies towards pro-active competition measures. In the US, for example, the General Accounting Office (GAO 1996) judged that the buy-sell market in slots established at certain US airports in 1985, had failed because it was argued to have done little to reduce slot concentration and introduce competition into the market. It went on to recommend that slots should be redistributed. In Europe, the current slot allocation regulation (Regulation 95/93, Article 10) requires that at slot coordinated airports[2] preference be given to new entrant carriers in the allocation of unused, returned or new slots (that together constitute the slot pool)[3]; there is also a reluctance on the part of the European Commission to accept unrestricted slot trading because of the fear that it will reinforce dominant positions; and approval of proposed alliances between carriers has often been accompanied by conditions requiring slot divestiture.

In this chapter, I suggest that these positions require re-examination; it develops a number of issues raised in Starkie (2003a). The effects of slot concentration are complex and there are a number of economic efficiency considerations that suggest that a (high) degree of concentration of slots at an airport might be beneficial. On the other hand, there are the understandable concerns that concentration, (that could come about by trading slots for example), could lead to reduced competition at the route level leading to a reduction in welfare. This tension between the advantages and disadvantages of slot concentration at airports

---

1   First published as 'The Dilemma of Slot Concentration at Network Hubs', in Czerny, et al. (eds) (2008).

2   A listing of coordinated airports in the EU will be found at Annex 2 of Czerny et al. (eds) (2008).

3   The European Commission has commented that 'new entrants [should be given] the possibility to grow [at coordinated airports] so as to reach a critical mass to be considered as viable competitors against the "big" carriers'.

has long been recognised and commented upon in previous papers.[4] Where this chapter is different is that it incorporates recent developments in the theory of congestion pricing that, it is suggested, have an important bearing on the balance of the argument. It also argues that higher fares at slot constrained airports do not constitute a *prime facie* case that a dominant position is being exploited and that there are a number of possible reasons apart from market power why we might expect higher air ticket prices at major hubs. Consequently, because the balance between the costs and benefits of concentration are difficult to determine, there should be a presumption against intervention in the market.

The next two sections outline the economic factors favouring slot concentration. This is followed by a consideration of the possible effects of concentration on consumer prices. A concluding section considers the implications for policy.

Networks and Natural Monopoly

Characteristics of Hub Networks

Most major and, as a consequence, often congested airports are hubs for an interconnecting network of air services and it is well known that (two-way) service networks connected through hubs produce powerful connectivity externalities (Oum and Tretheway 1995). In the case of two-way networks, as the number of nodes attached to the network increases so do the number of possible connections and thus its value to the consumer: the number of possible connections increases by the square of the number of connections (minus one). There is, therefore, the potential for a significant scaling effect so that the consumer benefits from an exponential growth in the number of possible destinations (s)he can fly to.[5]

In addition to increasing connectivity between nodes, the other important effect of a hub network is to increases the traffic density across the connecting flows and this also benefits the passenger by allowing for increased service frequencies. On the supply-side, an increase in traffic density is also associated with empirically established economies (Caves, Christensen and Tretheway 1984). Increasing the flow of traffic on a particular route for example, allows fixed station costs associated with the node to be spread over more passengers thereby reducing unit costs.

On the other hand, there are offsetting factors. For those passengers who, in the absence of a hub network, would have had the opportunity of a direct

---

4   See, for example, Kahn (1993) and Borenstein (1989). Borenstein comments: 'Though the link between airport dominance and high fares seems clear, a welfare analysis of increased airport concentration must also include the benefits that may accrue from hub operations ... These possible benefits ... should be weighed against the higher prices that seem likely to result' (362).

5   This is qualified when the nodes differ in their traffic generation potential. As Krugman (1999) has pointed out (referenced in Miller 2003), connections are usually made to the larger cities (nodes) first so that in practice, as more nodes are connected to the hub, the scaling effect decreases.

service to some of their preferred destinations (and no services to others), there is a price to be paid; longer journey times (but probably less schedule delay) and the inconvenience of having to change flights. Overall, however, there is a presumption that both passengers and airlines benefit from a business model based on hub networks. The issue here, however, is not the size of the overall net benefit; it is whether the characteristics of the hub network impart a degree of natural monopoly so that single firm delivery is more efficient. If this is the case, *ceteris paribus*, greater slot concentration at the hub is also more efficient.

Hub Networks as Natural Monopolies

The connectivity externality associated with hub networks is not in itself an argument in favour of service delivery by the single firm. To gain the connectivity externality the passenger does not require each connecting service to be operated by a single airline or alliance; the connectivity of the network could in theory be established with each hub connecting service operated by a different airline with slots allocated accordingly.[6] Equally, there is no obvious gain if the station costs of all the network *nodes* feeding through a hub are the responsibility of the single airline because there are no evident economies from combining these station costs. Consequently, the traffic density effect at route level would be captured equally well if each separate route in the hub network was operated by a different carrier.

Nevertheless, there is a demand and supply side gain to be achieved by single firm delivery across the whole of the network servicing a hub. In the case of the former, the benefit is from an improved quality of service. This improvement in service quality is the result of two factors. First, passengers have a preference for making *on*line connections rather than interlining between carriers, so that as the proportion of possible online connections grows, service quality increases (Bailey and Liu 1995). Second, service quality also improves if connecting times between services are minimised. Arguably, such times are more likely to be minimised for the maximum number of passengers, the larger the proportion of services provided by the singular airline. This is because of information asymmetries: to minimise total connection times in a non-concentrated network requires each airline to be fully informed of the flow dynamics on other parts of the network that it does not operate and also to act cooperatively on the basis of such information.[7]

---

6   This is facilitated by the IATA interlining system.

7   An interesting example of this problem and attempts to address it, has arisen in relation to the forthcoming opening of Terminal 5 at London Heathrow airport. The movement of British Airways into T5 will release space in two of the existing four terminals leading to an opportunity for a general reallocation of airlines to terminals. To assist in this task, the airport operator, BAA, commissioned work to assess what allocation would minimise connect times, taking into account flows of interlining passengers and the preferences of alliance partners to be co-located (Economics-Plus 1995). The analysis was based on the interlining data then available but inevitably failed to capture the dynamics of interlining.

On the supply-side, until recently empirical studies had suggested that there were no evident production economies from increasing the size of networks so that there were no efficiency gains if a singular airline providing services across all the routes in a particular hub-based network.[8] However, the more recent empirical literature on airline economics has indicated that, in addition to economies of traffic density, there are economies of network scale possibly associated with the better utilisation of aircraft fleets (Ng and Seabright 2001; Brueckner and Zhang 2001).[9] In addition, Jara-Diaz, Cortes and Ponce (2001) have suggested that previous approaches to measuring the effects of scale were confounded because an increase in network size adds more products (points) to the network; measurement, therefore, should take into account economies of spatial scope. Initial analysis on these lines using Canadian airline data suggests that there might be pronounced spatial scope economies, especially for the expansion of smaller networks (Basso and Jara-Diaz 2005). With finite slot capacity at a hub this cost advantage can be obtained only by slot concentration.

Thus, to summarise, there are considered to be net gains to passengers from networking services through an airport hub. Although these connecting services could be provided by as many airlines as there are routes serving the hub, the quality of the network is improved if it is possible for passengers to make online connections and, if these connections are organised within the firm, they are more likely to be optimised. This, in turn, suggests that service quality increases the greater the proportion of nodes linked by one carrier (or, but to a lesser degree of quality, an alliance of carriers); in turn, with finite airport capacity, this implies slot concentration can increase consumer welfare. Larger networks operated by the single carrier from a hub are also likely to introduce supply-side economies of scale or scope (possibly associated with improved equipment utilisation and with better utilisation of fixed establishment costs associated with the hub itself), again suggesting a degree of natural monopoly and thus efficiency gains from slot concentration.

Congestion Externalities

No Quantity Controls

At airports that act as network hubs, peak period congestion is a common feature so that an additional service imposes at the margin costs of delay on existing flight operations. The traditional prescription in these circumstances is to achieve an efficient level of use (and allocation of slots) by introducing (short run) marginal cost pricing to internalise the delay externality. Recent economics literature on the cost of airport congestion has, however, led to a major revision in thinking

---

8   See, for example, Gillen, Oum and Tretheway (1990) and Caves, Christensen, Tretheway and Windle (1987).

9   There might also be scale effects associated with establishment costs at the hub.

about the airport marginal cost issue and this new approach also has interesting connotations for how we might view slot concentration.

The new literature argues that the traditional approach to congestion externalities at airports has the effect of overstating the marginal costs of delay (Brueckner 2002, 2005 and Mayer and Sinai 2003). This is because the traditional approach does not take into account the fact that the delay costs that an airline imposes (by choice) on its own operations when it adds an additional flight to the peak at a particular airport are, in effect, internalised and therefore do not constitute an externality. The process of internalisation takes into account the impact of rescheduling on the operating costs of all other flights that the airline operates at the airport and, through adjustments to its peak/off-peak fares, the impact on the travel times of its passengers. (The latter are longer in the peak because of the added delay and, as a consequence, the airline has to adjust its fares to get the marginal passenger to switch to the peak; this required fare reduction affects all peak passengers thus reduces peak revenues as the airline adds flights. Equilibrium is reached when the gains and losses from shifting flights to the peak cancel each other). It is only with respect to the costs added to the flight operations of other airlines that the externality arises and therefore the marginal cost charge should reflect only this latter.

There are a number of consequences that follow from this insight.

- The traditional measure of marginal delay costs only applies to the airport case where, in the extreme, each service is operated by a different airline (the 'atomistic' case).
- Where, in the opposing polar case, all services are provided by a monopoly airline, there are no delay externalities, even when the airport appears to be congested; the efficient congestion charge is zero.
- If, as is typically the case, the airport has a mix of carriers operating different peak period quantities, different carriers should pay different levels of congestion charge; generally carriers with a larger number of flights will pay less.
- The different levels of congestion charge will encourage slot concentration.
- More slot concentrated outcomes will have the effect of reducing total congestion externalities; more slot concentrated airports will have on average lower delay externalities (Starkie 2003b).

With Quantity Controls

This revised view of congestion externalities was developed in the context of airports where there are no administered slot controls and airlines are free to add peak capacity by scheduling extra flights; the context is essentially that of a queuing model (see Daniel 1995). However, at a few US airports, together with most of the significant airports in Europe, the number of slots that can be used on an hourly basis is subject to a restriction, so that airlines have, in effect, to pre-book access; the object is to reduce the congestion externality by administrative means. This raises the issue of whether the revised conceptual approach to congestion externalities requires modification, or ceases to apply at all, at such slot controlled airports. This would clearly be the case, for example, if the administrative rules had the effect of eliminating congestion.

Generally, this is not the case; the controlled number of flight operations per hour is usually fixed at a level which is expected to result in some delay and it would be inefficient not to do so.[10] Of course, with the quantity of slots now fixed, an airline is no longer free to add to the *total* number of flights in the peak and therefore the equilibrium mechanism outlined above no longer applies in the same way, but some of the essential features of the new approach still remain. For example, if all peak flights at the quantity controlled airport were operated by the singular airline, the (residual) externalities associated with the expected delay allowed for in the slot control limit would be eliminated. Thus, the size of the externality is now a function of the distribution of the fixed number of slots and the more concentrated the distribution the smaller the overall externality. Consequently, at slot controlled airports rules that prevent the consolidation of slots, such as disallowing slot trading, might have the effect of increasing the total size of the delay externality.

Concentration and Prices

The objective of regulators interventions in the allocation of slots at major airports is to enable competitive entry and, in the case of the EU, this object is reflected in the preferences that must be given to entrants when allocating new slots. But in exercising this preference there is an opportunity cost involved; the benefit foregone by not allowing the incumbent network carriers access to the prescribed slots (unless potential new entrants decline their option). These benefits, as suggested above, include reduced congestion externalities, the higher quality of service to passengers associated with online transfers and the potential for lower costs as a result of density, scale and scope economies at the level of the firm. The riposte would be, however, that the potentially lower costs of the larger and more densely used network do not translate into a *reduction* in prices paid by airline passengers; in spite of potential economies of density, scale and scope, a number of studies have shown that fare yields on average are higher rather than lower, at airports where there are large incumbent carriers, with the implication that large carriers are exploiting their dominance.[11]

There are, however, a number of points to consider before it can be concluded that the higher fare yields at hubs do indicate an exploitation of market power. First, providing for transfer passengers including their baggage adds complexity and thus costs to airline operations and with more transfer passengers there is a

---

10   At London Heathrow, for example, the agreed average delay is 5 minutes but, in practice, the variance around this mean is considerable.

11   The GAO (1999), for example, pointed out that there were 13 airports serving large communities where passengers in 1998 paid, on average, over 8 per cent more than the national average fare. Seven of these airports were hubs for major airlines.

greater incentive to use expensive air-bridges rather than non-contact stands in order to speed up transfers. Second, hubs are usually large airports and larger airports are generally more costly to operate from because of the spatial separation of activities. For example, taxiing distances can be greater. Third, the organisation structure of a hub can add to airline costs because of the ground staff requirements associated with bunching arrival and departure flights to minimise connecting times. Fourth, the temporal concentration of on-line transfer opportunities also adds to demands for extra (peak) capacity which might, and should, be reflected in higher airport charges or in terminal lease fees if, as seems likely with major airports, there are diseconomies of scale in capacity.[12]

A number of factors, therefore, suggest that there are some operational costs that might be higher for airlines operating network hubs from major airports. Whether the factors pushing up operating costs offset the economies of density/scale/scope associated with a connected network is a moot point. If fares are higher from concentrated hubs, a basic issue is whether and to what extent this is a cost reflective outcome so that passengers are, in effect, paying higher prices for a differentiated and, ultimately, more costly quality of service, one that is associated with higher service frequencies and on-line connections.

There is a further important consideration. Many hub airports are increasingly operating at or near capacity, either operational capacity or declared capacity, particularly during peak periods. In these circumstances, prices (or quality deterioration) will have to be used to limit demand to capacity available. If the airport is pricing efficiently such rationing prices will be reflected in higher (peak) period charges, hence in higher costs to the airlines and, in turn, in higher fares charged to passengers for travel at peak periods. But for various reasons congested airports often charge airlines inefficiently low prices.[13] Nevertheless, it does not make sense for the airlines to pass on suboptimally low airport charges in the form of lower fares. If they were to do so, service quality would deteriorate (a growing number of frustrated customers would be unable to obtain a booking at posted prices). The sensible airline will maintain its fares at market clearing levels. This will result in high fare yields;[14] but these high yields will reflect scarcity and not monopoly rents.

There are, therefore, a number of good reasons why we might expect higher fare yields at airports dominated by hub carriers. The existence of higher yields *per se* does not indicate that dominant carriers are exploiting their market power. On the contrary, for airlines to extract true monopoly rents there has to be evidence

---

12   Starkie and Thompson (1985) argued from first principles that there were diseconomies of scale at major airports. Pels's (2000) empirical work supports this view with respect to terminal development.

13   See Starkie (2003b) for a discussion why airports seem reluctant to use efficient charging schemes.

14   For example, the GAO (1999) found that: 'Airfares at six gate constrained and four slot constrained airports were consistently higher than airfares at non-constrained airports that serve similar sized communities …' (20) but it tended to draw the conclusion that this must be due to exploitation of market power.

that output (as reflected either in the inefficient use of slots or in low aircraft load factors) is being deliberately restrained below capacity and the analytical work done so far on this point does not support such a conclusion (Kleit and Kobayashi 1996).[15]

Conversely, if you cannot increase output because of capacity limits, the proclivities of regulators to (re)allocate slots to potential competitors should make little difference to the overall level of fares at the hub; some fares may go down as capacity is shifted to allow competition on particular routes but, generally, this will be at the expense of other routes from which capacity is withdrawn and which as a consequence experience lower frequencies or the loss of a service. The effect of a reallocation of capacity in these circumstances is not to introduce competition across the network but, in large measure, to reassign the scarcity rents.[16]

Conclusions

This chapter has suggested that high levels of capacity utilisation by a single airline at major hub airports might constitute an efficient outcome and that recent developments in the theory of congestion pricing applied to airports has made this outcome more likely. The efficiency gains from concentration appear to have been largely ignored by policy makers who, instead, have focussed on the negative aspects of slot concentration, particularly on the reduced opportunities for potential service competition at the *route* level from a particular airport. In doing so, they have neglected competitive pressures arising in many regional markets from parallel or substitute routes facilitated by the use of secondary airports with spare capacity. These generally less crowded airports have been used by airlines with a business model different from that used by the typical network carrier. It is based on simplified low cost operations and, in Europe especially, point-to-point services.[17] Low cost airlines using this business model have had a major impact on local competition in the US domestic[18] and intra-European

---

15   Use-it-or-lose-it clauses in EU regulations and regulations governing the use of slots at high density airports in the US are also intended to guard against this practice.

16   At major congested airports there is rent-seeking behaviour. At the time of writing, the *Financial Times* (US edition, 3 November 2005) reports on the prospect of a deal between the EU and the US which could remove restrictions on further US carriers accessing London Heathrow. It comments: 'A deal would open London Heathrow to more competition (*sic*) ...'.

17   Although the low cost airlines have developed operational hubs, usually these do not provide for network traffic. Ryanair and easyJet for example, the two largest European low-cost carriers, provide for no connecting traffic at their respective hubs. Passengers can transfer between flights but have to purchase separate sector tickets. Ryanair has concentrated exclusively on what were, at the time of entry, secondary airports (although Dublin was an exception) and has made no use of the major network hubs, but easyJet has made some use of them [...].

18   See, for example, Morrison (2001).

markets[19] and, in the process, they have made rules such as that in the current EU slot Regulation giving preferential allocation of pooled slots to new entrants, unnecessary for the purpose of achieving competitive outcomes.

Thus, in judging an appropriate level of slot concentration there is an important balance to be struck between the benefits of concentration and the disbenefits associated with an increase in market power (increasingly tempered in short-haul markets by entry from low cost carriers using secondary airports). It is not at all clear where the optimal balance lies and this suggests a presumption against regulatory intervention especially interventions making use of *per se* rules. As matters stand it is by no means evident that the level of slot concentration to be found, for example, at major European network hubs (between 40 and 60 per cent of slots in the hands of the leading carrier), is inefficiently high once account has been taken of the inherent economies of density, scale and scope of network hubs, the reduction in congestion externalities of increased concentration and the growing competitive threat from low-cost carriers using secondary airports. The prevailing, often moderate, degree of slot concentration might, in fact, prove to be too low if efficiency and, thus, welfare is to be maximised.

---

19   See, for example, Davies et al. (2004).

References

Albon, R.P. and Kirby, M.P. (1983) 'Cost-Padding in Profit Regulated Firms', *Economic Record*, 59, 16–27

Armstrong, M., Cowan, S. and Vickers, J. (1994) *Regulatory Reform: Economic Analysis and British Experience*. MIT Press, Cambridge, MA

Armstrong, M. and Vickers, J. (1993) 'Price Discrimination, Competition and Regulation', *Journal of Industrial Economics*, 41, 335–59

Averch, H. and Johnson, L. (1962) 'Behaviour of the Firm under Regulatory Constraint', *American Economic Review*, 52, 1052–69

BAA (2003) Responsible Growth – *BAA's Response to the Government Consultation: The Future Development of Air Transport in the United Kingdom, South East and Eastern England*. BAA, London

Bailey, E.E. (1981) 'Contestability and the Design of Regulatory and Antitrust Policy', *American Economic Review*, 71, 178–83

—— and Liu, D. (1995) 'Airline Consolidation and Consumer Welfare', *Eastern Economic Journal*, Fall, 463–76

—— and Panzar, J.C. (1981) 'The Contestability of Airline Markets During the Transition to Deregulation', *Law and Contemporary Problems*, 44, 125–45

Barrett, S.D. (1984) *Airports for Sale: The Case for Competition*. Adam Smith Institute, London

Barrett, S.D. (2000) 'Airport Competition in the Deregulated European Aviation Market', *Journal of Air Transport Management*, 6, 13–27

Bartle, I. (2003) 'The UK Model of Utility Regulation', *CRI Proceedings 31*, CRI, Bath

Basso, L.J. and Jara-Diaz, S.R. (2005) 'Calculation of Economies of Spatial Scope from Transport Cost Functions with Aggregate Output with an Application to the Airline Industry', *Journal of Transport Economics and Policy*, 39, 25–52

Baumol, W.J. (1979) 'Quasi Permanence of Price Reductions: A Policy for Prevention of Predatory Pricing', *The Yale Law Journal*, 89, 1–26

—— (1982) 'Contestable Markets: An Uprising in the Theory of Industrial Structure', *American Economic Review*, 72, 1–15

——, Panzar, J.C. and Willig, R.D. (1983) 'Contestable Markets: An Uprising in the Theory of Industry Structure: Reply', *American Economic Review*, 73, 491–6

Beazley, K. (1983) Speech to the Aviation Law Association, October, Melbourne

Beesley, M.E. and S.C. Littlechild (1989) 'The Regulation of Privatised Monopolies in the United Kingdom', *Rand Journal of Economics*, 20, 454–71

Belobaba, P.P. and Van Acker, J. (1994) 'Airline Market Concentration: An Analysis of US Origin-Destination Markets', *Journal of Air Transport Management*, 1, 5–14

Better Regulation Task Force (2001) *Economic Regulators*, July, Better Regulation Task Force, London

Bishop, D. and Thompson, D. (1992) 'Peak Load Pricing in Aviation: the Case of Charter Airfares', *Journal of Transport Economics and Policy*, 16, 71–82

Borenstein, S. (1988) 'On the Efficiency of Competitive Markets for Operating Licences', *Quarterly Journal of Economics*, May, 357–85

—— (1989) 'Hubs and High Fares: Dominance and Market Power in the US Airline Industry', *RAND Journal of Economics*, 20, 344–65

Boyfield, K. (ed.) (2003) 'A Market in Airport Slots', *Readings, 56*, IEA, London

Brock, W.A. (1983) 'Contestable Markets and the Theory of Industry Structure: A Review Article', *Journal of Political Economy*, 91, 1055–66

Brueckner, J.K. (2002) 'Airport Congestion When Carriers Have Market Power', *American Economic Review*, 92, 1357–75

—— (2002) 'Internalisation of Airport Congestion', *Journal of Air Transport Management*, 8, 141–7

—— (2005) 'Internalisation of Airport Congestion: A Network Analysis', *International Journal of Industrial Organisation*, 23, 599–614

—— and Zhang, Y. (2001) 'Scheduling Decisions in an Airline Network', *Journal of Transport Economics and Policy*, 35, 195–222

Bureau of Transport Economics (1980) 'Basic Characteristics of General Aviation in Australia', *Occasional Paper* 33, AGPS, Canberra

—— (1981) 'Economic and Financial Issues Associated with General Aviation in Australia', *Occasional Paper* 34, AGPS, Canberra

Button, K.J. (ed.) (1991) *Airline Deregulation: International Experiences*. Fulton, London

—— (2002) 'Debunking Some Common Myths about Airport Hubs', *Journal of Air Transport Management*, 8, 177–88

Caves, D.W., Christensen, L.R. and Tretheway, M.W. (1984) 'Economies of Density Versus Economies of Scale: Why Trunk and Local Service Airline Costs Differ', *RAND Journal of Economics*, 15, 471–89

——, Christensen, L.R., Tretheway, M.W. and Windle, R. (1987) 'An Assessment of the Efficiency Effects of US Airline Deregulation via an International Comparison', in E.E. Bailey (ed.) *Public Regulation: New Perspectives on Institutions and Policies*. MIT Press, Cambridge, MA

Chamberlin, E H (1962) *The Theory of Monopolistic Competition*, 8th Edition. Harvard University Press, Cambridge, MA

Chancellor of the Exchequer (1978) *The Nationalised Industries*, Cmnd 7131, HMSO, London

Civil Aviation Authority (1991) BAA plc (*MMC2*). CAA, London

—— (1993) *Luton Airport Complaint against Stansted Airport: Decision*, APD7. CAA, London

—— (2002a) *Heathrow, Gatwick and Stansted Airports' Price Caps, 2003–2008: CAA Recommendations to the Competition Commission*. CAA, London

—— (2002b) *Economic Regulation of BAA London Airports (Heathrow, Gatwick and Stansted) 2003–2008, Proposals for Consultations*. CAA, London

—— (2003) *Economic Regulation of BAA London Airports (Heathrow, Gatwick and Stansted) 2003–2008, CAA Decision.* CAA, London

—— (2004a) *Airport Regulation, Looking to the Future – Learning from the Past.* CAA, London

—— (2004b) *Airport Regulation, Looking to the Future – Learning from the Past: Review of Responses to the CAA's Consultation Document.* CAA, London

—— (2005) A*irport Regulation: Price Control Review – Consultation on Policy Issues.* CAA, London

—— (2006) *Airport Regulation: Price Control Reviews – Initial Proposals for Heathrow, Gatwick and Stansted Airports,* December, CAA London

Competition Commission (2002a) *BAA plc: A Report on the Economic Regulation of the London Airports Companies,* CC, London

—— (2002b) *Manchester Airport PLC: A Report on the Economic Regulation of Manchester Airport PLC.* CC, London

Condie, S. (2004) 'Powerful Customers: Working with the Airlines', in P. Vass (ed.) *The Development of Airports Regulation – A Collection of Reviews.* CRI, Bath

Cornwall, N. (2004) 'International Trade in Gas and Prospects for UK Gas Supplies', *Lectures on Regulation, Series XIII.* IEA/London Business School

Czerny, A.I., Forsyth, P., Gillen, D and Niemeier, H.-M. (eds) (2008) *Airport Slots: International Experiences and Options for Reform.* Ashgate, Aldershot

Daniel, J.I. (1995) 'Congestion Pricing and Capacity of Large Hub Airports: A Bottleneck Model with Stochastic Queues', *Econometrica,* 63, 327–70

Davies, S., et al. (2004) 'The Benefits from Competition: Some Illustrative Cases', DTI, *Economic Paper 9.* DTI, London

Department of Trade and Industry (1998) *A Fair Deal for Consumers: Modernising the Framework for Utility Regulation.* DTI, London

Department for Transport (2003) *The Future of Air Transport,* Cm. 6046. DfT, London

Department of the Environment, Transport and the Regions (2000) *Appraisal Framework for Airports in South East and Eastern Regions of England.* DETR, London

Dixit, A. (1980) 'The Role of Investment in Entry Deterrence', *Economic Journal,* 90, 95–106

Doganis, R.S. (1992) *The Airport Business.* Routledge, London

Dresner, M., Windle, R. and Yuliang, L. (2002) 'Airport Barriers to Entry in the US', *Journal of Transport Economics and Policy,* 36, 389–405

Economics-Plus Limited (1995) 'Heathrow Terminal 5: Allocation of Capacity', *Report* to BAA plc. EPL, London

—— and GRA Inc. (2000) 'An Economic Assessment of the IATA Interline System', *Report* to IATA. Geneva, EPL, London

Findlay, C.C. (1983) 'Optimal Air Fares and Flight Frequency', *Journal of Transport Economics and Policy,* 17, 49–66

Forsyth, P. (1997) 'Price Regulation of Airports: Principles with Australian Applications', *Transportation Research – E,* 33, 297–309

—— (2001) 'Privatisation and Regulation of Australian and New Zealand Airports', *Journal of Air Transport Management*, 7, 19–28

—— (2004) 'Replacing Regulation: Airport Price Monitoring in Australia', in P. Forsyth, D. Gillen, A. Knorr, O. Mayer, H.-M. Niemeier and D. Starkie (eds) *The Economic Regulation of Airports*. Ashgate, Aldershot

—— (2006) 'Why Airlines Oppose Airport Peak Pricing', mimeo, August

——, Gillen, D. Mayer, O. and Niemeier, H.-M. (eds) (2005) *Competition Versus Predation in Aviation Markets*. Ashgate, Aldershot

——, Gillen, D. and Niemeier, H.-M. (2007) 'The Political Economy of European Airport Reform', paper to Airneth Annual Conference, Den Haag

—— and Hocking, R.D. (1978) 'Optimal Airline Fares, Service Quality and Congestion', *Forum Papers*, 4th Annual Meeting Australian Transport Research Forum, Perth

Foster, C.D. (1992) *Privatization, Public Ownership and the Regulation of Natural Monopoly*. Blackwell, Oxford

Fuhr, J. and Beckers, T. (2006) 'Vertical Governance between Airlines and Airports – A Transaction Cost Analysis', *Review of Network Economics*, 5, 386–412

Gillen, D., Oum, T.H. and Tretheway, M. (1990) 'Airlines Cost Structure and Policy Implications', *Journal of Transport Economics and Policy*, 24, 9–34

—— and Morrison, W. (2004) 'Airport Pricing, Financing and Policy: Report to National Transportation Act Review Committee', in P. Forsyth, D. Gillen, A. Knorr, O. Mayer, H.-M. Niemeier and D. Starkie (eds) *The Economic Regulation of Airports*. Ashgate, Aldershot

Graham, A. (2001) *Managing Airports: an International Perspective*. Butterworth Heinemann, Oxford

Graham, D.R., Kaplan, D.P. and Sibley, D.S. (1983) 'Efficiency and Competition in the Airline Industry', *Bell Journal of Economics*, 14, 118–38

Greenhut, M.L., Norman, G. and Hung, C. (1987) *The Economics of Imperfect Competition: A Spatial Approach*. Cambridge University Press, Cambridge

Grether, D.M., Isaac, R.M. and Plott, C.R. (1981) 'The Allocation of Landing Rights by Unanimity among Competitors', *American Economic Review*, 71, 2, 166–71

Giulietti, M. and Waddams Price, C. (2005) 'Incentive Regulation and Efficient Pricing', *Annals of Public and Cooperative Economics*, 76, 121–49

Hanlon, P. (1996) *Global Airlines: Competition in a Transnational Industry*. Butterworth-Heinemann, Oxford

Hatch, J.H. (1979) 'The Cost Minimising Number of Firms and the Discipline of Competition', *Economics Letters*, 4, 379–84

—— (1983) 'Contestable Markets: Their Relation to Other Recent Developments in the Theory of Competition and Market Structure', mimeo, University of Adelaide

Helm, D. and Thompson, D. (1991) 'Privatised Infrastructure and Incentives to Investment', *Journal of Transport Economics and Policy*, 15, 231–46

Hendriks, N. and Andrew, D. (2004) 'Airport Regulation in the UK', in P. Forsyth, D. Gillen, A. Knorr, O. Mayer, H.-M. Niemeier and D. Starkie (eds) *The Economic Regulation of Airports*. Ashgate, Aldershot

Hill, R., Starkie, D. and Thompson, D. (1985) 'Some Comments on the Airports Policy White Paper', Institute for Fiscal Studies, *Working Paper no. 77*. IFS, London

Hogan, O., Feighan, K. and Starkie, D. (2002) 'Off-Peak Landing and Take-Off Charges and Aircraft Classification', *Report*, Commission for Aviation Regulation, Dublin (at www.aviationreg.ie)

House of Commons Transport Committee (2003) 'Aviation', *Sixth Report* of Session 2002–03, House of Commons, London

—— (2006) 'The Work of the Civil Aviation Authority', *13th Report* of Session 2005–06, House of Commons, London

IATA (2007) Economic Regulation: *Economics Briefings No .6*. IATA, Geneva

Ironmonger, D. (1972) *New Commodities and Consumer Behaviour*. Cambridge University Press, Cambridge

Jara-Diaz, S.R., Cortes, C. and Ponce, F. (2001) 'Number of Points Served and Economies of Spatial Scope in Transport Cost Functions', *Journal of Transport Economics and Policy*, 35, 327–41

Kahn, A.E. (1993) 'The Competitive Consequences of Hub Dominance: A Case Study', *Review of Industrial Organization*, 8, 381–405

Kay, J., Meyer, C. and Thompson D. (eds) (1986) *Privatisation and Regulation – the UK Experience*. Clarendon Press, Oxford,

Kleit, A. and Kobayashi, B. (1996) 'Market Failure or Market Efficiency? Evidence on Airport Slot Usage', in B. McMullen (ed.) *Research in Transportation Economics*. JAI Press, Connecticut

Kunz, M. (1999) 'Airport Regulation: The Policy Framework', in W. Pfähler, H.-M. Niemeier and O.G. Mayer (eds) *Airports and Air Traffic: Regulation, Privatisation and Competition*. Peter Lang, Frankfurt am Main

Littlechild, S.C. (1983) *Regulation of British Telecommunications Profitability*. London: Department of Industry

—— (2003) 'The Birth of RPI-x and Other Observations', in I. Bartle (ed.) *The UK Model of Utility Regulation*, CRI Proceedings. University of Bath, Bath

—— (2007) 'Beyond Regulation', in C. Robinson (ed.) *Utility Regulation and Competitive Markets*. Edward Elgar, Cheltenham

MacKenzie, D. (1991) 'The Bermuda Conference and Anglo-American Aviation Relations at the End of the Second World War', *Journal of Transport History* 12, 61–73

Mandel, B. (1999) 'Measuring Competition: Approaches for (De-)regulated Markets', in W. Pfähler, H.-M. Niemeier and O.G. Mayer (eds) *Airports and Air Traffic: Regulation, Privatisation and Competition*. Peter Lang, Frankfurt am Main

Mason, K.J. (2000) 'Marketing Low Cost Airline Service to Business Travellers', *Journal of Air Transport Management*, 7, 103–9

Mayer, C. and Sinai, T. (2002) 'Network Effects, Congestion Externalities and Air Traffic Delays: Or Why all Delays Are Not Created Equal', *Working Paper No. 8701*. National Bureau of Economic Research, Washington, DC

—— and Sinai, T. (2003) 'Network Effects, Congestion Externalities and Air Traffic Delays: Or Why all Delays Are Not Evil', *American Economic Review*, 93, 1194–215

McGowan, F. and Seabright, P. (1989) 'Deregulating European Airlines', *Economic Policy*, October, 283–344

Memorandum of Understanding (1983) *Memorandum of Understanding between the Government of the United States of America and the Government of the United Kingdom of Great Britain and Northern Ireland on Airport User Charges*, see Appendix G, Starkie and Thompson, 1985

Miller, R.C.B. (2003) *railway.com*. Institute of Economic Affairs, London

Monopolies and Mergers Commission (1985) *British Airports Authority: A Report on the Efficiency and Costs of, and the Service Provided by, the British Airports Authority in its Commercial Activities*, Cmnd 9644. HMSO, London

—— (1996) *Report on BAA*, MMC4, London

Morrison, S.A. (2001) 'Actual, Adjacent and Potential Competition: Estimating the Full Effect of Southwest Airlines', *Journal of Transport Economics and Policy*, 35, 239–56

—— and Winston, C. (1986) *The Economic Effects of Airline Deregulation*. Brookings Institute, Washington, DC

Ng, C.K. and Seabright, P. (2001) 'Competition, Privatisation and Productive Efficiency: Evidence from the Airline Industry', *Economic Journal*, 111, 591–619

Office of Fair Trading (2006) 'UK Airports: Report on the Market Study and Proposed Decision to Make a Market Investigation Reference', *OFT 882*, London

Oum, T.H., Wates Jr, W.G. and Young, J.S. (1992) 'Concepts of Price Elasticity of Transport Demand and Recent Empirical Estimates', *Journal of Transport Economics and Policy*, 6, 139–54

——, Zhang, A. and Zhang, Y. (1993) 'Inter-Firm Rivalry and Firm-Specific Price Elasticities in Deregulated Airline Markets', *Journal of Transport Economics and Policy*, 17, 171–92

Panzar, J.C. and Willig, R.D. (1977) 'Free Entry and the Sustainability of Natural Monopoly', *Bell Journal of Economics*, 8, 1–22

Pascoe, F. (1983) 'Many Would Suffer under Domestic Airline Deregulation', *Australian Transport*, May, 13–15

Pels, E. (2000) *Airport Economics and Policy: Efficiency, Competition and Interaction with Airlines*. University of Amsterdam/Tinbergen Institute

Primeaux, W.J. (1977) 'An Assessment of X-Efficiency Gained through Competition', *Review of Economics and Statistics*, 59, 105–8

Productivity Commission (2002) 'Price Regulation of Airport Services', *Report 19*. Productivity Commission, Melbourne

Pryke, R. (1991) 'American Deregulation and European Liberalisation', in D. Banister and K. Button (eds) *Transport in a Free Market Economy*. Macmillan, London

Richards, K. (1996) 'The Effects of Southwest Airlines on US Airline Markets', in B.S. McMullen, *Research in Transportation Economics*, 4, JAI Press, Connecticut

Rovizzi, L. and Thompson, D. (1992) 'The Regulation of Product Quality in the Public Utilities and the Citizens Charter', *Fiscal Studies* ,13, 74–95

Sealy, K.R. (1976) *Airport Strategy and Planning*, Oxford University Press, Oxford

Schwartz, M. and Reynolds, R.J. (1983) 'Contestable Markets: An Uprising in the Theory of Industry Structure: Comment', *American Economic Review*, 73, 488–90

Secretary of State for Transport (1985) *Airports Policy*, Cmnd. 9542. HMSO, London

—— (1986) written answer to parliamentary question, 10 March

SH&E (2006) *Capital Needs and Regulatory Oversight Arrangements: A Survey of European Airports*. ACI Europe

Sherman, R. and Visscher, M. (1982) 'Rate of Return Regulation and Two-part Tariffs', *The Quarterly Journal of Economics*, 97, 27–42

Sibley, D. (2000) 'Economic Analysis of the Civil Aviation Authority's Initial Incentive Regulation Proposal for the National Air Traffic Services', Civil Aviation Authority Paper (www.caa.co.uk/erg/ergdocs/Sibley_NATS.pdf)

Soames, T. (2000) 'Essential Facilities and Transport Infrastructure in the EU', in J. Preston, H. Lawston Smith and D. Starkie (eds) *Integrated Transport Policy: Implications for Regulation and Competition*. Ashgate, Aldershot

Starkie, D. (1984) 'Contestable Markets, Response Lags and the Costs of Adjusting Prices', mimeo, University of Adelaide

—— (1989) 'De-regulation: the Australian Experience', in K.J. Button and D. Swann (eds) *The Age of Regulatory Reform*, OUP, Oxford

—— (1991) 'Airport Investment: the Monopolies and Mergers Commission Report on BAA plc's South East Airports', *Journal of Transport Economics and Policy*, 25, 303–6

—— (1992) 'Slot Trading at United States Airports', a *Report* for the DG VII of the European Commission. City Publications, London

—— (1994a) 'Regulating Airports and Airlines', in M.E. Beesley (ed.) *Regulating Utilities*, 37–55. Institute of Economics Affairs, London

—— (1994b) 'Developments in Transport Policy: The US Market in Airport Slots', *Journal of Transport Economics and Policy*, 28, 325–9

—— (1999) 'The US Department of Transportation's Statement on Predatory Conduct in the Airline Industry: An Alternative Proposal', *European Competition Law Review*, 20, 5, 281–6

—— (2000) 'A New Deal for Airports?', in C. Robinson (ed.) *Regulating Utilities: New Issues, New Solutions*. Edward Elgar, Cheltenham

—— (2001) 'Reforming UK Airport Regulation', *Journal of Transport Economics and Policy*, 35, 119–35

—— (2002) 'Airport Regulation and Competition', *Journal of Air Transport Management*, 8, 63–72

—— (2003a) 'Peak Pricing and Airports', in D. Helm and D. Holt (eds) *Air Transport and Infrastructure: The Challenges Ahead*. Oxera Publications, Oxford

—— (2003b) 'The Economics of Secondary Markets for Airport Slots', in K. Boyfield (ed.) *A Market in Airport Slots*. Institute of Economic Affairs, London

—— (2004) 'Testing the Regulatory Model: The Expansion of Stansted Airport', *Fiscal Studies*, 25, 389–413

—— (2005) 'Airports Regulation', in P. Vass (ed.) *Regulatory Review 2004/2005*. CRI, University of Bath, Bath

—— (2006) 'Investment Incentives and Airport Regulation', *Utilities Policy*, 14, 262–5

—— and Starrs, M.M. (1984) 'Contestability and Sustainability in Regional Airline Markets', *Economic Record*, 60, 170, 274–83

—— and Thompson, D. (1985a) 'Privatising London's Airports', Institute for Fiscal Studies, *Report Series 16*, IFS, London

—— and Thompson, D. (1985b) 'The Airports' White Paper: Privatization and Regulation', *Fiscal Studies*, 6, 30–42

—— and Thompson, D. (1986a) 'Stansted: A Viable Investment?', *Fiscal Studies*, 7, 76–82

—— and Thompson, D. (1986b) 'Stansted: Privatization Pre-empted', *Public Money*, 5, 5

—— and Yarrow, G. (2000) 'The Single Till Approach to the Price Regulation of Airports', *Report* to CAA, London

Starrs, M.M. and Starkie, D. (1983) 'Unregulated Airline Markets: The South Australian Experience', *Forum Papers*, 8th Annual Meeting, Australian Transport Research Forum, Canberra

Stern, J. (2003) 'What the Littlechild Report Actually Said', in I. Bartle (ed.) *The UK Model of Utility Regulation*, CRI Proceedings. University of Bath, Bath

Tirole, J. (2006) *The Theory of Corporate Finance*. Princeton University Press, Princeton

Toms, M. (2003) 'Is Airport Regulation Fit for the Purpose?', in D. Helm and D. Holt (eds) *Air Transport, The Challenges Ahead*. Oxera Publications, Oxford

—— (2004) 'UK Regulation from the Perspective of BAA plc', in P. Forsyth, D. Gillen, A. Knorr, O. Mayer, H.-M. Niemeier and D. Starkie (eds) *The Economic Regulation of Airports*. Ashgate, Aldershot

Train, K.E. (1991) *Optimal Regulation: The Economic Theory of Natural Monopoly*. MIT Press, Cambridge, MA

Tretheway, M.W. and Oum, T.H. (1992) *Airline Economics: Foundations for Strategy and Policy*. CTS, University of British Columbia

Turvey, R. (2000) 'Infrastructure Access Pricing and Lumpy Investments', *Utilities Policy*, 9, 207–18

—— (2001) 'Introduction', in P Vass (ed.) *Regulatory Review 2000/2001*. CRI, University of Bath

United States Department of Transportation (DoT) (1998) *Docket No. OST-98-3713: Statement of the DoT's Enforcement Policy Regarding Unfair Exclusionary Conduct in the Air Transport Industry*

United States Department of Transportation (1990) *Secretary's Task Force on Competition in the US Domestic Airline Industry: Airports, Air Traffic Control and Related Concerns (Impact on Entry)*

United States General Accounting Office (1996) *Airline Deregulation: Barriers to Entry Continue to Limit Competition in Several Key Domestic Markets*, GAO/RCED-97-4, Report to US Senate. GAO, Washington DC

United States General Accounting Office (1999) *Airline Deregulation: Changes in Air Fares, Service Quality and Barriers to Entry*, GAO/RCED-99-92, Report to US Senate. GAO, Washington DC

Vickers, J. and Yarrow, G. (1988) *Privatisation: An Economic Analysis*. MIT Press, Cambridge, MA

Walters, A.A. (1978) 'Airports – an Economic Survey', *Journal of Transport Economics and Policy* 12, 125–60

Waterson, M. (1984) *Economic Theory of the Industry*. Cambridge University Press, Cambridge

Whitbread, M. (1971) 'A Framework for the Efficient Production of Airport Services', *Regional Studies*, 5, 121–34

Windle, R.J. (1991) 'The World's Airlines: A Cost and Productivity Comparison', *Journal of Transport Economics and Policy*, 25, 31–49

Wolf, H. (1999) 'Tackling Congestion and Environmental Problems' in W. Pfhaler, H.-M. Niemeier and O.G. Mayer (eds) *Airports and Air Traffic Regulation, Privatisation and Competition*. Peter Lang, Frankfurt am Main

Index